Firefighter Rescue & Survival

Firefighter Rescue & Survival

Richard Kolomay Robert Hoff

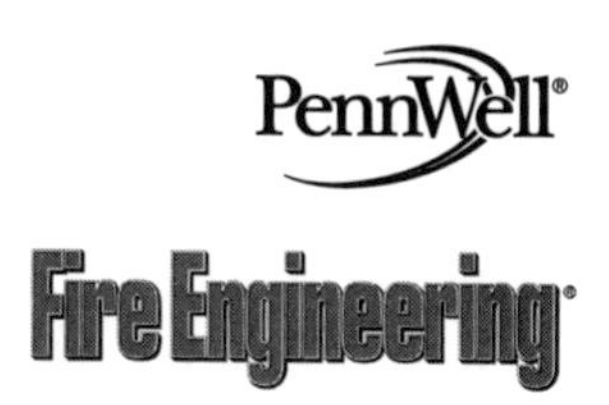

PennWell Corporation
1421 South Sheridan/P.O. Box 1260
Tulsa, Oklahoma 74112-6600 USA

800.752.9764
+1.918.831.9421
sales@pennwell.com
www.pennwell-store.com
www.pennwell.com

Cover design by Clark Bell
Book design by Amy Spehar
Supervising Editor: Jared Wicklund
Production Editor: Sue Rhodes Dodd

Library of Congress Cataloging-in-Publication Data

Kolomay, Richard.
Firefighter rescue & survival / by Richard Kolomay and Robert Hoff.--1st ed.
p. cm.
ISBN 0-87814-829-9
1. Lifesaving at fires. I. Title: Fire fighter rescue and survival.
II. Hoff, Robert (Bob) III. Title.
TH9402 .K65 2003
628.9'2--dc21
2003004880

Printed in the United States of America

2 3 4 5 07 06 05 04 03

Contents

chapter five

chapter six

chapter seven

chapter eight

chapter nine

Foreword

When Rick Kolomay and Bob Hoff first asked me to write this foreword for their book, I was moved to consider the state of firefighter rescue and survival training today…and reflect on how we've adjusted in the more than 32 years I've spent in the fire service. Indeed, we've made great strides. I remember the first working fire I ever fought, which was a serious cellar fire in a house in my hometown. During the course of the fire, a good friend of mine, Andy, was overcome with smoke while advancing the attack hoseline. Some hot wiring had melted through the low pressure breathing tube on his facepiece and cut off his air supply. The negative pressure mask was useless and he was now sucking in a lot of hot, nasty smoke. He made it as far as the base of the staircase before losing consciousness. Several chaotic minutes later, he was removed, unconscious but still alive, to the safety of the front lawn, where he was joined by several injured rescuers. "So, this is part of the job too?" I thought. "Besides rescuing civilians, we also rescue firemen. OK, if that's all there is to it, we can do that!" How simple those thoughts were in hindsight.

It was almost 20 years later, after having been involved in several more removals of unconscious firefighters, before the magnitude and nature of the skills required began to be recognized for what they are: among the most difficult tasks one can ever encounter on the fireground. I was working as a lieutenant in Rescue Company 2 in Brooklyn. We had removed unconscious firefighters up from cellars of factories and supermarkets, down from the floors above the fire in tenements, and out the window of a feather factory. All of these resulted in the survival of the unconscious members, but they all could have gone smoother. In one case, two of the rescuers were injured in the process. During our critiques of the operations, the members made many useful suggestions, then we went down into the firehouse cellar or to a vacant building and tried them out, with full gear, live victims, and blacked out facepieces. Most of the ideas worked, some did not. After a while, I began to think we were ready for anything. Then someone sent me a magazine article that is responsible for a tremendous change in the way firefighter rescue is approached today. It is not always a successful, almost routine effort as many of us were approaching the task. Instead, "The Murder of John Nance" showed the true nature of the task as a brutal, heart-wrenching affair, that may or may not be successful.

The article was first printed in The Columbus (Ohio) Monthly in December 1987, and was over two years old when I first read it in 1990. The description of the heroic rescue attempts at extreme personal risk by the members of the Columbus Fire Department sends chills up and down my spine every time I read it. Here was a group of men who went through hell trying to rescue a trapped brother but were unsuccessful. During my next tour at Rescue 2, we immediately set out to develop an action plan in case the same thing happened to us. (In fact, a very similar incident had occurred in Brooklyn 20 years earlier, when Lt. Jim Geraghty and another firefighter fell into a cellar full of burning tires and Lt. Richard Hamilton and Firefighter Lou Polera went down and hoisted them out.) With all the benefits of time, good visibility, no fire chasing us or low air supply, and with 20/20 hindsight, we set out to develop ways to remove firefighters from vertical holes, and we made sure that we would document them for the next guys.

Just as we were feeling pretty confident with these new techniques, another magazine article sent us back out to the vacant buildings. In September of 1992, Engineer Mark Langvardt of the Denver (Colorado) Fire Department became trapped behind a window gate on the 2nd floor of a commercial building. Again, despite Herculean efforts, the rescuers were not able to save one of their own. Pete Martin of Rescue 2 spent many hours developing methods for dealing with what is now referred to as "the Denver scenario." As these techniques were tried, each new group took the steps a little further. The material was spread by word of mouth from firehouse to firehouse. In New York City, it ended up being taught in the Technical Rescue School. In Illinois, Rick, Bob, and the staff at the Fire Service Institute started to include segments at a class on private dwelling fires during the annual Fire College. Before long, "Saving Our Own" was born out of the recognition that firefighter rescue is truly different from civilian rescue, and is a subject that requires a great deal of analysis, study, and practice. The course has shown its worth in the lives of the firefighters it has helped to save.

Back in New York, I had been promoted and moved on to be the Captain of Rescue Company 1, where every new man in the unit had to demonstrate that he could "rescue the Captain" from each of the scenarios. Out windows of vacant buildings, up through the 20" x 20" hatch in the roof of the rescue rig, up staircases, down staircases, and anything else you can name. The Rescue School was teaching our rescue company firefighters the methods to save lives. Again, that feeling of "being ready for anything" started to set in. Then in June of 1999, yet another tragedy showed the need for even greater dissemination and emphasis on the techniques presented in this book. Captain Vincent Fowler was overcome in the cellar of a small, private house fire. There was no rescue company on the scene. Again, despite extraordinary efforts, by the time Captain Fowler was removed from the building, he had absorbed a fatal dose of carbon monoxide and other poisons. On the day of Captain Fowler's funeral, Chief of Department Pete Ganci ordered the implementation of a department-wide program to train every member of the department in these techniques, so that, hopefully, we would never lose another firefighter who might otherwise have been saved. The message is clear. You can never feel like you can handle everything and anything. We all must strive every day to improve our skills and abilities. We cannot expect somebody else to save our friends and co-workers. It is up to each and every one of us to prepare ourselves and our units for this task.

As you read these techniques, try to put yourself in the place of those who went before you. Keep in mind the circumstances they were working under. It's not like it is on the drill ground or in the cellar of your firehouse, where you don't have to worry about the room "lighting up" around you as you drag the dummy across the floor or wonder if the floor you are dragging along is going to fall out from under you. As I said, every time I read "The Murder of John Nance" it sends chills up and down my spine because I can put myself in the position of guys like Tim Cave who were doing everything in their power to rescue their friend. However, you have an advantage over Tim. You have the benefit of his experience without having to live through it yourself. You have people like Bob Hoff and Rick Kolomay, who have spent years practicing and refining the techniques you will need to be successful in your mission. Many of you will discover new ways to resolve problems. That's great! This will be how we all learn. Take the lessons you learn here and share them with others. We owe it to John, Mark, Vinny, and all of the others.

Battalion Chief John Norman, FDNY, Chief in Charge, Special Operations Command

Acknowledgements

The material in this book is intended to be a reflection of the many dedicated and professional firefighters and fire instructors across the United States of America. As your contributions continue, the rescue techniques will improve and expand with time, but so will the rescues. Please stay united and strong.

From Rick to Patti, Brandon, Carissa, and Noelle Kolomay, I fully realize the time sacrificed from our home life was great. Thank you for your understanding and love. To my mother Catherine, who always encouraged reading, writing, and a good education throughout my life.

To my father, my mentor, Sully Kolomay. As a proud Chicago FD firefighter who had me at his side taking in fires when I was six on a 1954 Mack engine (Engine Co. No. 98), providing guidance and inspiration which can in part claim responsibility for this book.

From Bob to Joyce, thank you for your understanding of our mission with this book, for always being there for me, and showing me life beyond the fire service. You truly are an "angel." To my daughter Sara and my son Andy, thanks for all your support.

A personal dedication to my father, Chicago FD Assistant Drillmaster Thomas A. Hoff, who was killed in the line of duty on February 14, 1962, you are my inspiration.

To my mentor, and the best truck company officer, Chicago FD Battalion Chief Patrick Healy (deceased), who taught me from a very early age the true meaning of being a "fireman."

To Bennie Crane, this book would not have been possible without you! Your direction, motivation, and undaunted support was more than anyone could have asked for. As authors, and more so personal friends, we thank you from our hearts.

On a special note to Chicago F.F. Sully Kolomay, and posthumously Battalion Chief Tom Hoff Sr., in February of 1962 you both spoke with each other in the fire academy drill yard just days before Chief Hoff's life was lost in the line of duty at an apartment building fire. The inspiration and motivation that each of you have provided to your sons, exactly forty years later, will be forever encrypted on every page of this book.

To Battalion Chief Ken Wood, Schaumburg (IL) FD, we personally and professionally thank you for your time and great effort in reviewing, correcting, proofreading, and commenting on every single page of this book with the greatest of care and diligence.

To Roy Hervas, who was adopted as our official photographer, we thank you very much. His personal time and dedication for not only this project, but to the fire service overall is greatly admired and appreciated.

To Battalion Chief John Norman, FDNY SOC, back in 1987 when you were writing the first edition of the *Fire Officer's Handbook of Tactics* book, you ignited another lifelong goal to write a book. As a fire service leader and mentor and a personal friend, we thank you very much for your support and guidance. It will never be forgotten.

Chicago (IL) Fire Department: Battalion Chief Ray Hoff, District Chief Andy O'Donnell, Deputy District Chief Ray Orozco, Deputy District Chief Ed Enright, Lt. Will Trezeck, Capt. Tom Magliano, F.F. Brandon Dyer, Lt. Pat Lynch, F.F. Tom Meziere, Battalion Chief Rich Edgeworth, F.F. Brian McArdle, F.F. Gary Coney, Battalion Chief John McNamara, District Chief Ed Gavin, Lt. Cliff Gartner, Battalion Chief Bob McKee, F.F. Art Noonan, Capt. Richard Ford, F.F. Shane Haynes, Lt. Jim Altman, F.F. Corey Hojek, District Chief Cortez Holland, Capt. Mitch Crooker, Engineer Skip Straeder, CFD Chaplin Father Tom Mulcrone, and Fire Commissioner James T. Joyce.

Strong heartfelt gratitude goes to the many **Schaumburg (IL) Fire Department** firefighters for their time, effort, and sacrifice to help support this project. Special mention must include, Deputy Chief Terry Simale, Battalion Chief Rick Anderson, F.F. Jack Schneidwind, F.F. Jim Klausing, Lt. Fran O'Shea, Lt. John Steele, F.F. Scott Sutherland, F.F. Butch Adams, F.F. Marty Diaz, Lt. Bill Spencer, Capt. Mark Harper, F.F. Harry Seibert, F.F. Doug Tragesser, F.F. Brian Johnson, Lt. John Brohan, F.F. Mark Toussaint, and Lt. Marty Sand.

Great appreciation must be given to the **Illinois Fire Service Institute** and the Saving Our Own program, in particular Director (ret. U.S. Marine Col.) Richard Jaehne, Deputy Directors David Clark, Jack McCastland, and Jim Straseske (ret.) who provided us with the motivation and opportunity to instruct and challenge much of the fire service with firefighter rescue and survival training.

To the many dedicated and loyal fire service professionals from across the country who not only helped to contribute and support the efforts of this book, but also became part of a larger "family," we thank you very much:

Battalion Chief Don Hayde, FDNY, Lt. Sal Marchese, FDNY, Deputy Chief Craig Shelly, FDNY, Lt. Pete Lund, FDNY, F.F. Nick Giordano, FDNY, Capt. Fred Dimas Sr., Phoenix (AZ) FD, Lt. Mickey Conboy, FDNY, Battalion Chief John Salka, FDNY (Get Out Alive program), Battalion Chief Butch Cobb, Jersey City FD, F.F. Seth Dale, Darien-Woodridge FPD, F.F. Jay Comella, Oakland (CA) FD, Battalion Chief Ted Corporandy, San Francisco (CA) FD, Capt. Jack Rutledge, Peach Tree (GA) FD, Lt. Vinny Russell, Boston (MA) FD, Battalion Chief Rick Fritz, High Point (NC) FD, Harvey Eisner, Firehouse Magazine, Capt. John "Sully" Sullivan, Worcester (MA) FD, Lt. Kevin Malloney, Worcester (MA) FD, District Chief McNamee, Worcester (MA) FD, Asst. Chief TJ Welch, Alameda County (CA) FD, Battalion Chief Gerry Kohlmann, San Jose FD, Capt. Mike Spalding, Indianapolis (IN) FD, F.S. Instructor Joe Nedder, Massachusetts Firefighting Academy, Fire Chief Dick Schaeffer, Lawrence (MA) FD, Capt. John Hojek, Oak Lawn (IL) FD, Deputy Chief Tom Shervino, Oak Lawn (IL) FD, Deputy Chief Tom Brady, Aurora (IL) FD, Capt. Mike Clarke, Bath (ME) FD, Eileen Coglianese and the Chicago Gold Badge Society, Asst. Chief Diana Watkins, Marquette Heights (IL), Capt. Dave McGrail, Denver (CO) FD, Dr. Denise Smith PhD, Skidmore College NY, Asst. Chief Hugh Stott, West Chicago FD, Lt. Steve Wilcox, Wheaton (IL) FD, F.F. Michael Griffiths, Lake Zurich (IL) FD, F.F. Brian Herli, Chicago FD.

Special mention to F.F. Andy Fredricks, FDNY Squad 18, and Lt. Ed Coglianese, Chicago Engine 98, who have provided us with both spiritual guidance and motivation to work endlessly to help carry on their unfinished work. Thank you, and God bless.

Introduction

After traveling much of the United States and working intensely with fire departments large, small, career, combination paid, paid-on-call, and volunteer, we have found many of their concerns with firefighter safety to be similar. Their greatest concern is the rescue and survival of their own. Many potentially catastrophic incidents have been shared with us that never made the press or trade journals. The stories were both inspiring and emotionally moving. Many were near misses where firefighters were, for some reason, spared serious injury or even death; some were not near misses, but direct hits.

The mission of this book is to assist in the effort to reduce the number of line-of-duty (LODD) fire-fighting injuries and deaths, while at the same time being proactive in fire service training and leadership. The gravity of this mission can only be truly understood by those who have been missing or trapped and those who will be responsible for search and rescue of those who become missing or trapped. To the members of the fire service who have never assisted in the removal of a seriously injured firefighter or picked up the dead body of a firefighter to be placed in a body bag, **please always remember to take your fire-fighting profession seriously—never stop learning and be receptive to change.**

Deliberate training in firefighter rescue and survival is a field that is new to many in the fire service and private industry alike. It should be understood that much of the material in this book has been generated due to many fatalities of firefighters in the past and it is an unfortunate fact that there will be many more fatalities to contend with in the future. Since there will be more fireground fatalities, there will be new lessons learned, and therefore, it is believed that no book can fully and thoroughly cover the entire field of firefighter rescue and survival as is the case with this publication. What must be agreed upon in the field of firefighter rescue and survival is the statement, "Never give up!"

After this introduction was originally written and ready for submittal, our world and the fire service suddenly changed. This great nation was brutally attacked on September 11, 2001, resulting in the largest loss of life to the New York City Fire Department and the fire service in history. Altogether, 343 firefighters lost their lives at the World Trade Center. While trying to comprehend what happened and why it happened, this project was set aside until we could fully comprehend how our brother firefighters had been attacked and bombed in an act of war against the United States of America. The situation was that they had not fallen victim to peacetime fire-fighting operations. In spite of the events on September 11, 2001, we strongly decided that the mission of this book has not changed. What has changed is the meaning of

"Never give up!" It not only means to fight-to-survive in the context of firefighter rescue and survival but also to never give up in getting stronger as a fire service and as a great nation.

Rapid intervention during an emergency is the act of immediate involvement to assist, search, and/or rescue. A team of firefighters or specially trained personnel at the scene of the emergency in a combat-ready position with gear, equipment, and tools is called a rapid intervention crew or team (RIC or RIT). Much to the credit of the New York City Fire Department, the concept of a rapid intervention team was actually conceived after the Equitable Bank Building fire in Manhattan on January 9, 1912. This fire occurred during the early morning hours with extremely cold temperatures and winds up to 65 mph. As firefighters attempted to fight the fire all the way up to the fifth floor, the fire became out of control and spread floor-to-floor, thereby trapping and killing firefighters and civilians. During the fire, the president of the Mercantile Savings Deposit Company entered the cellar to retrieve the bank securities. When suddenly cut off by subsequent collapse and fire, the banker and a clerk were trapped behind a cellar window covered by bowed 2-in. iron security bars.

After 2½ hours of hand-sawing with a hack saw, the bars were finally bent apart enough by firefighters to free the two men for rescue and medical attention. The extended time and effort it took for this rescue resulted in the creation of a special rescue unit—Rescue Squad Company #1. This unit carried the most up-to-date rescue equipment and trained personnel and had the stated purpose to perform difficult and unusual rescues and to rescue firefighters who needed to be rescued.[1] Unfortunately, between the 1900s and 1980s, with the exception of New York City, much of the fire service did not formally address any concept or training of firefighter rescue.

The 1997 edition of the National Fire Protection Association (NFPA) Standard 1500, *Standard on Fire Department Occupational Safety and Health Program,* Section 6-5 (Appendix 1) addressed RITs, and it outlined a concerted national effort towards firefighter rescue. In an effort to comply with the NFPA, standard questions started to arise such as the following:

- Who should be the rapid intervention crew?
- Who would they report too?
- What would they do when they arrive?
- How should they respond if needed for rescue?
- Where do we get the people to make a rapid intervention team when we cannot get enough people to fight the fire? (This is the most frequently asked question by most fire departments.)

Our training experiences in suburban and rural America has revealed that there is no doubt that, in most cases, fire department resources are sparse at best. Many times, whatever resources might be available to staff the efforts to fight fire are not used due to outdated practices, leaders who are not informed, and ongoing border wars. One of the most controversial, yet supportive, set of standards for firefighters are NFPA 1710 and 1720, which address staffing recommendations for paid urban/suburban and paid on-call/volunteer fire departments. As mentioned in the *Fire Officers Handbook of Tactics* by FDNY Battalion Chief John Norman, "To rescue people or accomplish any other task, you must have able-bodied personnel available to do the job!" That is a given.

The fact is that a minimum staffing of two firefighters to compose a RIT is not only ineffective but also potentially dangerous. The removal a semiconscious or unconscious firefighter from a second floor window requires a minimum of four rescuers provided that surrounding conditions are stable. Committing two firefighters to such a rescue is like sending them on a Kamikaze mission. (This is a reference to Japanese pilots during WW II who purposely flew airplanes into enemy targets, knowing that they would die also.) The demands placed upon the RIT members when they are deployed to perform firefighter search and rescue are tremendous. Physical challenges are presented in the form of ladder raises, crawls, victim drags, and extrication. Equipment familiarity under the harshest of conditions is necessary to operate thermal imaging cameras, supply self-contained breathing apparatus (SCBA) air, and work with search and rescue rope. Possibly the most challenging demand is the mental toughness that is required to enter a building in uncontrolled fire conditions, with a potential for collapse, along with the added knowledge that a time bomb is ticking relative to the survivability of one of our own. Firefighter distress situations, basically, do not occur when efforts to fight fire are going well. Bluntly put, when things "go to hell," the RIT is needed most.

Some of the most common responses by some fire officers and firefighters alike when assigned the position of rapid intervention are:

"You have got to be kidding me, RIT!"

"What? You want us to just stand here?"

"Chief, if you get us out of RIT, we'll put this fire out!"

"Who's idea was this?"

"Ya sure, we'll stand here as the RIT or Rectal Insertion of Thumb."

"We waited a year for a fire, and now we get to watch it from the front yard."

"Leave the RIT for the 'Out-standing firefighters'!"

"We can be the RIT for rehab!"

"Wow, we get to herd sheep in the front lawn again!"

In spite of these protests, it should be remembered that to be in a trained state of readiness, prepared for a catastrophic event that might endanger, injure, or cause death to our brothers and sisters who are committed to fighting fire, is an extremely noble cause. For those who do realize the responsibility of the RIT, the perseverance to fulfill the role is ongoing until ordered to "stand down." They are company officers and firefighters who are mature and experienced, having fought a number of fires, who now realize they must prove their courage and skill in another way.

The aggressiveness of RIT members is shown by their ability to communicate with the incident commander (IC) and other team members concerning structural and fire conditions while preparing for "worst-case scenarios." Additionally, they must execute the immediate staging of tools in accordance to policy and conduct a size-up of the fireground.

Focus and discipline are two essential behaviors for any RIT officer. Keeping the team fully informed and involved in the RIT operation is critical. It might involve the staging of additional tools, additional size-up, or an update of scene accountability. Even when there is "down time" on the fireground, the team must be kept informed and kept together as a team. Realizing that complacency can be a killer, the RIT officer again must stay focused.

One of the most stressful commands for an IC to make is to send a RIT into a building to rescue firefighters. They could be entering an unstable structure with an uncontrolled fire condition and with the emotion that one or more of their own is missing and/or trapped. The fact is that if there is a collapse or a Mayday distress call is heard via radio, the "weight of the world" will be placed on the RIT in a split second. They must be mentally and physically prepared as well as fully equipped. If any of the ten quotes mentioned previously were seriously recited when assuming the responsibility of the RIT, not only will they not be prepared, but they can become victims themselves. The event is mixed with adrenaline, fear, emotion, and pure dependency upon one another as a team to get the job done.

Hence, there comes a time when we must provide guardianship for those who are committed to saving lives and property, and now is the time to realize the responsibility and importance of that guardianship. As written by the late John F. Kennedy in *Profiles in Courage,* "A man does what he must, in spite of personal consequences, in spite of obstacles and dangers, and pressures—and that is the basis of all human mortality."

1

Proactive Firefighting Tactics and Training

Injury and Death Statistics

The first chapter of this book is explicitly dedicated to those who have suffered the supreme sacrifice of duty. As positive as the firefighter rescue and survival training may be, it still cannot be forgotten that the unfortunate reasons for such training are the case studies and experiences from those who have been heroically injured and killed in the line of duty. A very profound quotation by retired FDNY Deputy Chief Vincent Dunn explains this chapter best: "**In order for a firefighter to survive the danger of firefighting, he or she must know how other firefighters have died.**"

On-Duty Firefighter Fatalities (1977–2001)

Year	Deaths	Year	Deaths
2001	441	1988	136
2000	102	1987	131
1999	112	1986	121
1998	91	1985	126
1997	94	1984	119
1996	95	1983	113
1995	96	1982	125
1994	104	1981	135
1993	77	1980	140
1992	75	1979	126
1991	109	1978	171
1990	108	1977	157
1989	119		

As objectively put as possible, statistics[1] from the year 2000 show that there were 1,708,000 fires attended to by public fire departments in the United States and that it was the lowest recorded number of annual fires recorded by the U.S. Fire Administration. As a result, one would assume that firefighter line of duty deaths (LODD) also declined. However, this is not the case. They have remained constant and in 1999 actually worsened as a single year. From 1990 to 2000, firefighter deaths averaged 106 per year (one LODD every 3½ days), and firefighter fireground injuries averaged 55,509 per year (152 injuries every day). As encouraging as some years have been with decreased LODDs, there have been other years that have offset the decreased years, leaving an upward spike in statistics.

These alarming statistics have prompted in-depth investigations from the National Institute for Occupational Safety and Health (NIOSH) with the goal of preventing severe injury and death and providing an excellent resource for learning. NIOSH[2] was funded in 1998 to investigate firefighter fatalities throughout the United States. The goals of the NIOSH LODD investigations are as follows:

- To better define the magnitude and characteristics of work-related deaths and severe injuries among firefighters
- To develop recommendations for the prevention of these injuries and deaths
- To implement and disseminate prevention efforts

As the NIOSH firefighter Fatality Database has supplied information for this book, it will also supply valuable information for local fire service health and safety programs and training programs as well. Although NIOSH explicitly expresses that their investigations are to call national attention to injuries and fatalities of firefighters and not to place blame or fault, it must be recognized that research reveals many fireground injuries and fatalities are due to the following repeated reasons:

- Poor physical fitness
- Poor fire/rescue training
- Inadequate staffing
- Poor communication and communication systems
- Inadequate equipment and apparatus such as SCBA (self-contained breathing apparatus, PASS (personal alerting safety system alarms), and thermal imaging camera
- Lack of written fireground procedures
- No accountability system

However, there are also firefighter injuries and fatalities that can be classified as non-preventable, which may result from an unpredictable explosion (e.g., hidden propane), structural collapse (e.g., unknown building modification), a rapid change in fire behavior (e.g., high winds), or other similar situations. Such non-preventable LODD catastrophes will only stop if the fire service stops routinely entering buildings for suppression and search or just does not respond at all. Neither option is realistic and for that matter, the end result could undoubtedly be far more catastrophic.

Based on a National Fire Protection Association report[3] in 2002, it was declared that after a 24-year study of on-duty firefighter deaths from 1977 to 2000, firefighters today are dying inside structure fires at a rate that parallels the 1970s. In spite of the declining rate of structure fires and the improvements in personal protective equipment, SCBA training, communications, and incident command, the number of firefighter fatalities in structure fires has not comparably decreased.

As updated training indicates, structural components from engineered lumber, as well as lightweight steel and aluminum, are dominating newly built structures throughout the world. With such lightweight construction, there are significantly shortened time-tolerances for firefighters to operate inside such buildings when the structural components are being attacked by heat or direct flame. For that matter, we are also victimized by older vintage buildings that are defying gravity and "standing by habit" as opposed to structural strength. When ravaged by heat or direct flame, they fall apart at the seams in a catastrophic manner taking down entire floors, walls, roofs, and anything or anybody in or around them. Realizing that structural collapse is the leading fireground killer, the proactive measures that need to be taken by every firefighter are essential. Fireground strategy training sessions, building construction classes, pre-plan tours, and self-education through books, video, and CD are all avenues that can be used to increase the awareness of a firefighter's "Number One Killer."

It is hoped the combined efforts of sound leadership, quality training, and lessons learned from the fatality reports in conjunction with strong enforcement of a rapid intervention team (RIT) policy, that a definitive and constant decline in firefighter injuries and fatalities will result. The fact is if we did not address proactive measures to prevent firefighters from getting into catastrophic situations, we would not be laying a solid foundation for any of the material found in this book.

Proactive Fire Service Leadership

Part of the proactivity to be discussed in relation to firefighter rescue and rapid intervention operations is the craft of leadership. Rapid intervention experiences and training to date have proven that very strong leadership must prevail in the fire station and on the fireground. When a firefighter sends a Mayday distress call and fire conditions are worsening, a company officer will have to move the RIT into position. Emotions are high, fear is present, confidence is in question, and yet a firefighter must be rescued. It is a time when a company officer must recognize the risks of collapse, the behavior of smoke, and the effects of increasing levels of heat. They must keep account of the team, sustain radio communication, and most of all maintain a "rock-steady" sense of command so the team will follow and perform as needed. As indicated earlier, structural collapse has been responsible for claiming more firefighter lives on the fireground than any other cause. Therefore, once the initial collapse has occurred and firefighters are trapped, the RIT officer must be even more alert to and knowledgeable about the weakened structure and the possibility of a secondary collapse. What if that officer is not trained or experienced enough to recognize a primary collapse potential and the resultant dangers? The question then arises, "Should that individual be the RIT company officer?" Unfortunately, most fire departments do not have the resources or staffing to be selective as to who becomes the officer. For that matter, many fire departments across the United States are struggling just to assemble the necessary staffing to organize a RIT.

One of the most recognized problems in the present fire service is the lack of company officer training at the most basic level. With this in mind, we are now demanding company officers in charge of RITs to operate at an even more advanced and intense level than ever experienced. For those firefighters and company officers who are assigned to a RIT unit, not making the correct split-second decisions, such as immediately recognizing changes in fire behavior or failing to evaluate their level of SCBA air, can result in the loss of the lives of the entire team. This being the case, the fire service has to rise to the occasion by increasing command and control training for all members of the fire service.

A fireground company officer must be aware of and continually monitor the following items:

- Potential collapse
- Current fire conditions
- Changing smoke conditions

Additionally, the RIT officer must avoid the following behaviors:

- Issuing vague, uncertain, and incomplete orders to personnel on the scene
- Failing to make any tactical decisions
- Failing to maintain accountability for assigned personnel
- Failing to maintain good radio communications (lack of a remote microphone at the collar)
- Exhibiting a lack of confidence and initiative
- Failing to exhibit a "presence of command" while operating on the fireground
- Failing to call for help in a timely manner or not at all

During basic engine company fire attacks or truck company operations (non-rapid intervention situations), if the company officer fails in any of these areas, the result could be disastrous for all firefighters involved. In many cases though, during normal firefighting operations, there is a supportive team effort with an incident commander (IC) and several company officers to accomplish a coordinated fire attack and search operation. By operating as a team in this manner, if an officer falls short with a "presence of command" or in any other area, the strengths of other members and companies will work together for a successful fire attack. Even though this is not the most desirable scenario, it helps to equalize the fire attack results.

However, this is not true for RIT operations. **It has been discovered that during rapid intervention operations, there is very little tolerance allowed for any deficiency or weakness in a company or chief officer.** Although each firefighter rescue mission is unique, many of the same negative dynamics exist:

- Confusion
- Anxiety and panic
- Denial
- Retreat

The heightened level of emotions combined with the need for immediate action in firefighter rescue situations leaves no room for error or delayed decision-making by a RIT officer or IC.

Confusion

As it is, a working structure is a small-scale disaster; with any disaster comes the element of confusion. Confusion may exist among the occupants, neighbors, and in some cases the responding firefighters due to the severity of the fire. Compound this original level of confusion with a structural collapse, flashover, or explosion, and confusion can turn to complete chaos. It is this level of chaos that a RIT officer must navigate through to accomplish the mission. What has been proven to help sort out such intense confusion and chaos is an officer who is alert, aware, and ready with a plan of action. Sometimes there is no stopping the "wave" of firefighters from moving into the affected area where firefighter victims are trapped, but coming through that wave will be an officer and a team who have a plan of action, proper equipment, and training. It will be up to the chief officer(s) to gain control of the scene as soon as possible to allow the RIT to operate as needed.

Anxiety and panic

We are, first of all, human beings, then rescuers. As much as we pride ourselves on being able to maintain courage in the face of danger, certain natural instincts of protection take over when our lives are in danger. Such instincts might manifest as adrenaline-driven excitement, yelling, illogical thought, involuntary movement, running, and many other actions. For instance, at a live-burn training session, after a planned flashover demonstration, a veteran firefighter crawled out the front door uninjured, stood up, immediately ran across the street in full protective clothing never to be seen again, and days later resigned from the fire department. There is no doubt that despite the high level of safety precautions and warnings prior to the flashover, this individual felt highly endangered and reacted with panic. During an actual fire attack or rapid intervention operation, anyone who falls into a panic mode may be subject to a fatal outcome. A RIT officer who has planned for the worst and trained on possible scenarios will be much more capable of suppressing any feelings of panic.

Denial

Again, as we are human beings first, the emotion of denial can overwhelm any rescuer to the point of total "shutdown." A very effective way to describe the emotion of denial would be to ask the reader to reflect back to one of the most devastating events of his or her life and recall his or her initial reaction to the situation. Many may have reacted in silence, some may not have been able to believe what they were being told and requested the information to be repeated again and again, and some may have just sat down and wept. There have been many chief officers who simply shut down from the effects of denial after a catastrophic event wherein firefighters were missing or known to be killed. Since emotions can be controlled, the RIT officer will have the advantage of preparing for and preplanning worst-case scenarios. In the event of a catastrophe it will be easier to deal with the reality of the event and react appropriately without any overwhelming sense of denial or without experiencing shutdown.

Retreat

The act of retreating from a burning building should not be because a firefighter signals a Mayday in the belief that they are missing, lost, or trapped. In the event of a

Mayday distress call, any retreat of interior attack companies should be based on collapse hazards, sudden change in fire conditions, or a change in fireground strategy to support members of the RIT. Fire fighting companies should NOT be evacuated from a burning building for such things as role calls or personal accountability reports. This could result in the abandonment of critical hose streams, which may knowingly, or unknowingly, be protecting the endangered firefighters. Additionally, it could stop an effective search and rescue operation.

Each of these four emotions can affect all firefighters, company officers, and chief officers without bias. It is, therefore, critical that the RIT officer is not affected to the point of hampering the mission of the RIT. The training, experience, and aggressive preparation of the RIT will determine, in most situations, how effective the team will be under the worst of conditions when activated to go to work.

Company and chief officers commanding RIT operations should carefully observe the following crucial advice:

- Show a confident "presence of command"
- Make every action thought-out and deliberate
- Communicate slowly and clearly to all members of the RIT and command officers
- Verify any communication transmitted or received, and use a remote microphone
- Maintain complete accountability by keeping the RIT together
- Never say never; never say always
- Physically lead the team at all times and be assertive

Show a confident "presence of command"

Just as the acts of a RIT may be unusual or unethical according to normal firefighting standards, the leadership of the team can be too. As best defined by Army General Colin Powell, "Leadership is the art of accomplishing more than the science of management says is possible." "Presence of command" is an image and a feeling that is witnessed by subordinates. It is unique to each individual leader and is projected by an officer during times of adversity when lives are in danger. A good example is the late actor John Wayne who had a very natural and unique image of "command" that was capitalized on by Hollywood filmmakers. Its not to suggest that everyone should view the movie "The Green Berets" before every tour of duty, but the seriousness of projecting a confident image is very important. Such images and feelings are conveyed in many different ways. Very different from that of John Wayne is retired Army General Colin Powell who commands with a lower profile but maintains a very stern, strong, direct, and competent "presence of command." What helps greatly with an individual's "presence of command" is the person's quality of reputation and amount of credibility within the fire

Fig. 1–1 Chicago Lieutenant and Later Schaumburg (IL) Fire Chief, Robert Sutherland Demonstrates a Strong Presence of Command As He Prepares to Search for Three Missing Firefighters

service. Basically, no team of level-headed, sane firefighters would follow an officer into "hell" when the officer is fumbling around with an SCBA strap, mumbling, and looking away from the scene as if to be praying for help before committing to a RIT operation.

As seen in Figure 1–1, Chicago F.D. Lieutenant Robert Sutherland assigned to Snorkel Squad #1 and later Fire Chief of the Schaumburg (IL) F.D., had a very strong "presence of command." As in this photo, Lt. Sutherland had just been informed of three missing firefighters, and he was preparing to enter for a search.

When activated for duty, RIT officers will experience fear. As bad as fear can be, it can also be positive. Fear raises the natural senses to an all-time high. The sense of sight, sound, and touch in particular become extra sensitive to changes such as heat conditions, smoke behavior, structural shifting, falling ceiling, breaking glass, sounds from victims, and verbal and radio communications. To an experienced officer, this extra sensitivity to the conditions and the situation is almost subconscious and instantaneous and can be described as a "gut feeling." Such a feeling has been defined by some as a hollowed feeling in the stomach filled with fear. Others have felt twisting and stabbing pains in the stomach, requiring several deep breathes to continue moving forward. Many officers and firefighters alike have shared their experiences with "gut feelings" and how they managed to survive a near miss with the grim reaper because they followed their "gut feelings."

Such instincts and feelings are very important for an officer, especially a RIT officer, to have for a presence of command. FDNY Battalion Chief Don Hayde described a "gut feeling" of retreat during a situation when he was a captain in Brooklyn. While on the second floor of an ordinary-constructed building with heavy fire below and in the walls along with heavy smoke conditions, he recognized the large number of firefighters with him. His "gut feeling" was to issue the order to begin backing out. Captain Hayde had instinctively surmised that the building might be weakening when, suddenly, the floor dropped about 6 inches. Fortunately, everyone retreated safely, and there were only minor burns. His "gut feeling" was on target. The only personal disappointment that he shares was the fact that he did not act on that feeling sooner and back the companies out before the floor dropped. This, in itself, is a lesson for us all. Although a "gut feeling" might not be the most scientific term to relate the concept of fight-or-flight during an emergency, the reality of the matter still comes down to the following equation:

Learned instincts through experience +
several fearful incidents +
proactive training =

A dependable "gut feeling."

Make every action thought-out and deliberate

This is easier said than done, but it is ever so critical. This is where experience as an officer counts in terms of being able to think under fire. As far as RIT operations are concerned, one must be able to think under fire without the interference of emotion as we attempt to rescue our own.

Communicate slowly and clearly to all members of the RIT and command officers

The number one problem at any routine fire is generally communication, whether it be radio communication, hardware problems, or verbal interpretation. It is important for the RIT officer(s) to maintain composure and communicate slowly and clearly. Miscommunication during a RIT operation can result in disaster, which has been already proven numerous times.

Verify any communication transmitted or received, and use a remote microphone

The importance of redundant radio and face-to-face communication is imperative. In one case, while initial companies were arriving at a fast-spreading fire in a combined church and school building (over 100 years old) on a cold winter's night, an interior attack occurred through the front door. As the first engine advanced through the church toward the fire in back of the altar, the volume of overhead fire and in the back of the building was unknown to them. As secondary companies arrived, they deployed, according to policy, in the rear setting up a tower ladder. The secondary companies began to report via radio that heavy fire was coming from the windows and roof and that the tower ladder was going to flow water. Unfortunately, the initial engine company did not receive that vital radio message before the roof began to collapse separating the firefighter on the nozzle position from the rest of the company. Though most were able to escape through the front doors, the firefighter on the nozzle position was lost and died in the building.

What lessons did the fire service learn?

1. The tower ladder company who transmitted the heavy fire conditions should have verified that the IC received that information. It was the determining factor as to whether the fireground strategy should have gone to a defensive mode thereby immediately retreating any companies' performing interior operations.
2. Every portable should have a remote microphone attachment. This microphone should be affixed to the firefighter's collar, on a radio strap, or positioned in any other innovative way so it will be close to both the ear and mouth for a "hands free" operation where it can be heard and quickly spoken into. Although the IC had a remote microphone, it was clipped to the radio antenna. When the IC completed a transmission, the radio (and remote microphone) was physically lowered to his waist or used as a pointer, which did not allow any transmissions to be heard. In this case, the ever so important transmission from the tower ladder company was not heard, and the engine company inside was not pulled out in time.

Fig. 1-2 The Remote Microphone Positioned Near the Collar (Hervas)

Maintain complete accountability by keeping the RIT together

Let safety and common sense prevail. After numerous reviews of RIT case studies, actual incidents, and training experiences, it has been found that the RIT operates most effectively and safely staying together as a team based on the following ratio:

RIT rescue 4 to 1

Whether the RIT is arriving on the scene with four or more personnel or assembling the personnel on the fireground, it has been found that the

Fig. 1–3 The RIT Operating as a Cohesive, Prepared, Trained, and Strongly Led Group (Kolomay)

RIT should not split up into separate search teams when deployed for a Mayday call. The most common reason for separating a RIT would be to search for multiple firefighter victims, but the chances are great of the RIT failing their search and rescue mission when split up. It must be remembered that the RIT company officer is highly dependent upon each team member for rope management, emergency SCBA air, tools, and manpower. Separating the RIT could mean the loss of the search rope, emergency SCBA air, and the possible loss of team accountability during dangerous, deteriorating fire conditions. Reinforcement of RIT cohesiveness is important. A RIT company must understand the importance of calling for additional help to reduce the need of splitting up. One of the most important mission statements of the RIT is to be part of the solution not add to the problem.

There is also the reality that many fire departments who serve rural and some suburban areas do not have adequate staffing to attack the fire, much less to stage a rapid intervention with at least four personnel. For those fire departments who cannot supply enough staffing for a RIT, it will be important to assign a team of two firefighters as soon as possible to carry out the duties of an efficient RIT. If needed to deploy to a Mayday distress call, that team of two must be able to immediately recruit on-scene firefighters to assemble a fully staffed RIT. The disadvantages of having to recruit on-scene personnel are:

1. **Fatigue.** They have been fighting the fire and now might have to expend an additional 110% of their energy to rescue a firefighter. It is more likely that those spent firefighters can become victims themselves rather than those who can respond specifically as a RIT.
2. **SCBA.** Depending on the amount of time into the fireground operations, recruited firefighters for the RIT may have half or less air available in their SCBA.
3. **Review of scenarios.** Recruited firefighters might be familiar with the interior of the building, but they will not have had time to mentally prepare for a possible Mayday scenario and the various rescue methods that will need to be used.

4. **Accountability**. In the confusion and rush of recruiting on-scene firefighters to become part of the RIT, those personnel who are assigned might not be accounted for by the incident commander (IC). In some cases, certain personnel might assign themselves to the RIT without authority to do so, which places them in a freelance mode without accountability.

Fire departments who have difficulty staffing RITs will have to consider mutual aid. Because of the need for rapid intervention as early as possible into the fire's duration and the long travel distances of some mutual aid fire departments, it is important for ICs to call for mutual aid as soon as possible. Some fire departments have designed an automatic mutual aid system that responds by sending personnel upon the confirmation of a working fire. In any case, it must be realized that wherever the staffing comes from, it still requires an average of four firefighters to search and rescue one firefighter victim.

Never say never; never say always

There will be a time when the RIT can use a trained and planned "divide and conquer" type of procedure. If the RIT must resort to a reconnaissance mission before committing to any type of all-out search and rescue operation, it will be necessary at times for the RIT officer and a firefighter to advance slowly on a search line with a thermal imaging camera and appropriate hand tools to determine both structural and fire conditions (this will be covered in detail in subsequent chapters). Although the reconnaissance will only go so far into the building with a divided team, it is a trained and planned procedure within the scope of RIT operations. There may also be a time when there will be multiple firefighters and possibly even civilian victims right next to each other, requiring the RIT to subdivide. Another situation might require one or two RIT personnel to stay with a trapped firefighter to provide shared SCBA air and moral support until extrication efforts can begin.

These are some of the examples explaining how and why the RIT can become divided, while maintaining accountability. We hesitate to say "never" and "always" because the fireground, as well as many other emergencies, do not always allow for a clear-cut textbook procedure, especially when it comes to firefighter rescue techniques.

Physically lead the team at all times and be assertive

Generally, when a RIT enters the building, it is at a point when both the tactics and the structure are going from bad to worse. It has been found that a weak officer, one who does not assert orders, follow up on progress, and organize search and rescue efforts in a strong manner will ultimately fail the RIT operation. With RIT operations taking place in deteriorating fire conditions, the RIT officer must be visible with assertiveness and a "presence of command" without giving way to excitement or panic. The officer must keep accountability, control, and communication among the RIT members. Although there may be moments when the RIT officer has to leave the RIT firefighters to give or receive direction from a window or doorway, it can only be momentary. The RIT officer MUST be present to instill calm and order at all times. At any time when there is doubt in the mind of a rescuer due to fatigue, fear, difficulty in the rescue, low air, increasingly hot and/or smoky conditions, and many other reasons, they will (and should) look to the officer for guidance.

Fig. 1–4 Company Officer Provides Strong Direction and Leadership of Firefighters Operating at a Single-Family House Fire (Hervas)

Proactive Company Officer Training

As long as we still commit our personnel to interior searches for people and perform interior fire attacks to suppress fire, this profession will remain a blue collar, dirty, hands-on occupation requiring training to be of the same likeness. This is true, not only for normal firefighting and rescue operations, but especially for RIT operations, which present an even greater demand for preparedness and skills. Basic training with self-contained breathing apparatus, ladders, ropes and knots, search procedures, and even physical fitness take on new value when it can be demonstrated that the skills learned may also be used in the rescue of fellow firefighters or even ourselves. In addition to basic training, the cross-training of all personnel is very important. Cross-training on various types of apparatus such as tower ladders, aerial towers, and aerial ladders should be a priority. If a RIT is assigned to an engine company or a mutual aid department and they must use the platform from the tower ladder to remove a firefighter from a window or roof, they have to be familiar with how to "put it to work."

The need for this proactive training is the same for proactive leadership because it is for the firefighter. The complexity of training can vary from the use of a sledgehammer to organic chemistry. Whatever the topic, the company officer must know how to maintain a balance between the "basics" and the "specialized."

Fig. 1-5 The Company Officer Committed to Participating and Leading the Training (Kolomay)

Street-wise Training Hints for Command Officers

1. Keep your training sessions brief—10 to 30 minutes unless the firefighters demand more.
2. Always attempt to mix academic with "hands-on."
3. Relate the training to its context. Stay realistic!
4. Require that each firefighter participate at some point during the training session.
5. Present the opportunity for challenge, not embarrassment.
6. Focus on behavior rather than personality. It is the appropriate action, not attitude, that will extinguish the fire or succeed in a rescue.
7. Train at the company level in small groups of four (4) to eight (8) individuals in order to maintain involvement and interest.
8. Design training for results, not processes. Enable the firefighters with the training and education needed to accomplish their goal safely and effectively without getting bogged down in procedures. Firefighting is about one result—getting the fire extinguished.
9. As long as the training objective has been achieved, let the group run with it!
10. It's okay to have fun, too.

Training can involve setting up a dry hoseline for a simulated basement fire in a firehouse for live-burn training. A very good example of taking training out of the classroom and into the field occurred in a major city where we were training on firefighter rescue. We had stretched some hose in an effort to demonstrate some basic engine operations with firefighter and officer positioning while committing to an interior attack. The failure of this basic engine operation had resulted in a firefighter fatality in Chicago. While in full turnout gear on the hose line and demonstrating how the engine officer should be positioned behind the nozzle position, a deputy chief in the class requested that we repeat our demonstration. Specifically, he asked that we repeat the demonstration and emphasize that the officer was to be positioned behind the firefighter on the nozzle during the fire attack. At that point we knew "something was up." After honoring his request, he explained that a number of officers assigned to engine companies in his department had been taking control of the nozzle instead of allowing the firefighters to do so. The reason he stated was that it was a "fun" position to be in and the officers had the authority to set up their engine operations as they saw fit. As he continued, he pointed out that when the officer is more concerned with having fun, they cannot effectively watch the fire conditions around them, listen to radio communication, or advance line effectively or safely, as was also the reason for the demonstration. Had this basic "hands-on" training not been brought to our attention or had it been left for "chalk talk," it is clear that this valuable critique, and any subsequent changes it brought about in their engine operations policy, would not have occurred in the Chicago Fire Department. District Chief Bennie Crane made a most profound statement concerning training that we will never forget and always live by, "If they [firefighters] can't say it and they can't show it, then they don't know it."

If the officer is not versed on a particular subject area, then choose someone in your company or department who is versed in that area and able to conduct the training. If necessary, the department may have to reach outside (e.g., state instructor, private sector, police, and many other agencies) to gain the training and knowledge needed.

Having said all of this, **the most valuable ingredient for fire service training for the company officer is support from the top of the organization.** This support consists of the four (4) following sub-components:

1. Financial

As quoted in the movie *The Right Stuff,* "No bucks, no Buck Rogers!" The monetary support to train firefighters to be firefighters and officers to be officers is essential. It is difficult to translate training that will save a firefighter's life into dollars and cents, but unfortunately, it is a fact of life. Depending on the type of fire department organization, more or less funding is still needed for equipment, training props, training aids, payroll, and tuition. In many cases, without money, the training cannot and will not happen.

2. Morale

A sense of mission must be provided as to how the fire department is expected to perform at a working fire by establishing the needed levels of aggressiveness, experience, versatility, and safety through training. Like motivation, morale comes from within the individual firefighter, and it is a feeling that varies from person to person. In most cases, morale is highest after an actual incident at which individuals, companies, and the department perform efficiently and safely. Training is the next best tool to raise morale. Subsequently, firefighter rescue training has provided a much more meaningful training experience, since those whom we are rescuing are generally familiar to us with recognizable names and faces.

3. Experience

Providing opportunities for firefighters and officers to mentally and physically become proficient within the scope of their rank not only helps them learn the standard operating procedures, but it also teaches how to deal with and react to the "gray areas." Each and every fireground fatality scenario has been different with unique circumstances in terms of weather conditions, structure type, building modifications, cause and origin of fire, the fireground tactics, staffing and experience levels, and countless other variables. Such variables have and will require the officer to use unconventional training techniques and in some cases, special tools to search for and rescue a firefighter in distress. As discussed earlier, commanding officers must take their normal duties of command and control to a higher level during RIT operations. This requires not only knowledge, but also field experience. Experience allows an officer to "reach back" to know what works and what does not work, aside from what the books say. This takes training, time, and experiences (both good and bad).

4. Participation

When possible, the greatest support of all is when a chief officer can participate in the training as well. The ability to lead by example is one of the most necessary attributes in the fire service at this time. If a chief officer can take the opportunity to "roll around in mud" with the troops, then the leadership, energy, respect, and morale will skyrocket in the eyes of many.

If chief officers set a course to improve or maintain company officer interest, aggressiveness, and discipline in the field, then their energy and support for quality training will be priceless. Obviously, not every company officer or firefighter can be proficient at every task, but as a team, we can. According to Lee Buck's Second Deadly Sin of Leadership:

> *"Every act of conscious learning requires the willingness to suffer an injury to ones self esteem. That is why young children before they are aware of their self-importance learn so easily, and why older persons who are vain or self-important have great difficulty in learning. Pride and vanity can thus be greater obstacles to learning than stupidity."*

Perhaps we do risk our pride when we have to ask to "say it" or "show it" during training, but we also minimize the real risk—the risk of an injury or fatality when confronted with the same situation in a non-training situation. It is also the duty of a company officer to train, and even more so, our obligation to other firefighters and their families.

2

Rapid Intervention Operations

Combat-Ready

Initially, the opinions of persons from various areas of the fire service were that rapid intervention operations should be a non-combative position until needed. The assumption was that assigned RIT members would stand, with tools in-hand, next to the command post until activated. However, the rapid intervention position has actually developed into a very active and combative position. Due to the potential expectations of a RIT, their activity has been found to be constant while working on behalf of the firefighters inside the building. Not only will physical activity be taking place, such as tool staging and size-up, but also mental alertness in keeping up with the ever changing incident. Similar to military operations, "combat-ready" refers to a team of members who are equipped, skilled, and prepared to do what needs to be done to ensure an effective and safe rescue of an endangered firefighter.

One of the most frequently occurring difficulties during rapid intervention operations is keeping the discipline among the "troops" as they get into position for their RIT duties. As companies around them are stretching hoselines or raising ladders, it is easy to set down the RIT tools and help out. Let common sense prevail, but do not forget the mission! The RIT officer may often be bombarded with requests from the RIT firefighters for opportunities to check the rear of the building or to take a look inside. Although their requests might be valid, it has actually occurred that the RIT officer was the only one left in the staging area. No matter what the RIT is doing while staging, the only question that must be answered at any given time is, "Are we combat-ready to deploy now?" If not, that is the time for the officer to immediately regroup and brief the team.

Risk vs. Benefit

The concept of *risk versus benefit* during normal fire fighting operations is viewed with respect to firefighter safety. During rapid intervention operations, a textbook view of risk versus benefit would be no different, but because rapid intervention operations are anything but textbook operations, so are the risks during a Mayday distress call. There may be skeptics who would argue that statement, but let us set experience aside and define the balance of risk versus benefit in both normal fire fighting operations and rapid intervention operations. Such risks in both operations can be identified as:

- Structural conditions
- Fire conditions
- Hazardous materials
- Rescue operations

The rapid decision that must be made by both chief officers and company officers inside and outside of the building to determine whether a company of firefighters will carry on with a dangerous maneuver is based on the following factors:

- Training
- Past experience
- Urgency of the rescue operations
- Potential for structural collapse
- Amount and speed of fire spread to other properties
- Exposure to hazardous materials
- Potential for explosion

Then, given normal fire fighting operations, the question now is "What benefits are to be gained by committing fire fighting personnel into a certain tactical operation under certain fire fighting conditions?" The following is a list of possible benefits.

- Search and rescue for possible occupants
- Rescue of known occupants
- Confinement of a fast-spreading fire
- Roof ventilation
- Forcible entry
- Water evacuation and salvage
- Overhaul
- Locate and rescue missing, lost, and trapped firefighters

As discussed in chapter 1, proactive fire service leadership and the anticipated emotions that these rapid intervention officers must deal with (confusion, anxiety/panic, denial, and the need to retreat) will likely result in a much higher level of risk. Putting a RIT to work is as risky as it is necessary. In addition to all the factors already mentioned, there are even more risks that every IC in particular must be fully aware of.

The urgency to save

The urgency to save a firefighter often will engage an adrenaline drive that will alter or even block out sound decision-making when fireground officers evaluate the risks. Other factors that influence such an urgent drive upon rescuing firefighters can involve high levels of anxiety, fear, and external loss of control causing uncontrolled yelling and radio traffic.

Collapse potential

If the building suffered a partial collapse, the chance of additional collapse is almost certain. In this case, a RIT can become victimized for the same reason the original victim(s) had become trapped, thus adding to the disaster.

Fire conditions from "bad to worse"

At a time when the efforts to fight the fire are "losing" the building and there is a Mayday distress call, the fire conditions will go from bad to worse. When a Mayday distress call is heard, many of the fire fighting efforts will tend to stop and gravitate toward the rescue effort. The result is a rescue attempt with an out-of-control fire, very limited time, extremely high risk, fireground errors, and a diminished chance of finding those in trouble. The study of one fatality case indicated that when fire companies in the rear of the building were notified of a firefighter down in the front on the second floor, they abandoned their positions to assist with the rescue but did not turn off the positive pressure fans in the rear. As a result, the fire continued pushing into the rescue area as those on the hoselines tried to fight the fire back and away from the rescue operation. Unfortunately, fireground errors and the inherent risks of fire fighting are a reality and a fact that we must recognize during a Mayday situation.

Limited time to affect the rescue

Heavy fire and smoke conditions requiring rapid intervention members to use the SCBA throughout the search and rescue operation will limit time by three fireground factors:

1. Fire conditions
2. Collapse potential
3. SCBA air consumption

Understanding that even if the firefighters can gain the upper hand by confining the fire to buy time, the RIT is still under the time constraint of their SCBA. The fire service generally operates with 30-minute air cylinders that average 20 minutes of actual use. Providing the RIT members have full SCBA cylinders, it is essential that the firefighter rescue be accomplished within that 20-minute time span. During deteriorating fire conditions, no matter how simple or complicated the rescue may be, if it is not accomplished by the initial RIT before their SCBA air runs too low, their success will be doubtful and the risk becomes very high. The difficulties of trying to supplement SCBA, change air cylinders, exchange entire SCBAs, or use shared-air techniques is difficult, dangerous, confusing, and time consuming.

Leaving the firefighter victim

Leaving the firefighter victim will be another complication in the rescue operation. Due to the limited time that the RIT will have, if they do not remove the victim within the time span of that first cylinder of SCBA air, they will be in a position to have to leave the victim to obtain more air or be replaced by another RIT. Experience has shown that these actions are easier said than done. Once again, having to consider the emotional and adrenaline drives of the rescuers many complications can arise:

Complication #1. The RIT entering the building might have to contend with the victim's partner, officer, or other firefighters who are low on SCBA air or have no air left, but who do not want to leave. Depending on the smoke and heat conditions inside, this reluctance to leave can now easily cause them to become victims, thus compounding the Mayday situation.

Complication #2. Some incidents will be difficult for the RIT to just get into the rescue area because existing firefighters will not physically move to allow the RIT into position to go to work. In many cases, the existing firefighters will have already started some type of rescue effort. Their rescue effort might be successful or might fail due to entanglement, entrapment, or just a loss of direction resulting in mass chaos. It is hoped that once the RIT officer identifies who they are, existing firefighters will fall back on past rapid intervention training to work with the RIT or evacuate as ordered.

Complication #3. A loss of direction and focus can quickly occur among both existing firefighters and the RIT during a rescue. If leadership, communication, or composure among the rescuers breaks down, so will the rescue operation. Yelling, arguments, and even physical disagreements result, none of which are in the best interest of the victim. This complication is not a criticism—it is a warning for what has and will happen again. One of the concepts of the RIT is not to become part of the loss of direction but to be the source of new direction for communication, leadership, and composure to perform a successful rescue of a fallen firefighter.

Complication #4. If the original rescue efforts required a more specialized rescue involving air bags, hydraulic tools, or rope lifting techniques, and the rescue operation must be handed over to another RIT, the transition is very difficult at best. Deteriorating fire conditions, falling debris, confined work spaces, poor communication, and the time it will take to inform the next RIT of what has been done and what still needs to be done, can make the situation nearly impossible. If feasible, it might require the RIT officers to meet outside of the rescue area, such as at a window or doorway, before committing another team. The obvious advantage would be a "team briefing" that consists of clear and concise information about the operation before the new RIT enters heavy smoke conditions with their SCBA gear on.

Unlike normal fire fighting operations, firefighter rescue operations have a very clear, concise, and urgent benefit—to locate and rescue one of our own! Experience and the many case studies involving firefighter fireground fatalities have demonstrated consistently how the risk versus benefit scale tips quickly in favor of the victim when a RIT is activated.

RIT Staging

The IC successfully stages the RIT based on their direct visual and verbal contact with the RIT. This is a working relationship similar to the one between the IC's sector officers and company officers. The dynamics of the RIT staging area are:

Identification and recognition

The IC can identify who the RIT officer is, and the RIT officer can identify who the IC is. The reality of that relationship is knowing the strengths and weaknesses of each other. Most importantly are the strengths of both being:

1. Presence of command
2. Experience
3. Crisis management skills
4. Awareness of time and tactics
5. Communication using both radio and verbal skills

Fig. 2-1 The RIT Officer Has the Ability to Watch the Command Post or Sector Officer and Constantly Monitor the Fire Scene Progress (Hervas)

Visual contact

The IC can visually see the RIT. The IC or sector officer can see that the RIT is combat-ready with tools prepared in staging. Some RITs can loose discipline and virtually stray from the staging area or may otherwise need to be deployed.

Monitor

The RIT officer has the ability to watch the command post or sector officer and constantly monitor the fire scene progress. Closely watching the IC and/or sector officer and observing their behaviors, decision-making efforts, issuance of orders, level of calmness, confidence, and control will provide the RIT officer with a great amount of information concerning the incident. As so wisely stated before, "we are, first of all, human beings then firefighters," due to that fact we are subject to forgetfulness, confusion, physical illness, frustration, distraction, or "just having a bad day." For these reasons and perhaps many others, the person in charge can simply make a mistake or fall behind. Unfortunately, those errors have and will cost the firefighters their lives. This is even more reason to observe those who are responsible for scene control because they may the be first indication that things are getting out of control.

Fig. 2–2 The RIT Officer Works with the Sector Officer When Possible to Assist in Determining the Condition of the Structure and Firefighters Operating in the Context of Firefighter Rescue (Hervas)

Fireground education

The RIT officer should take time to instruct firefighters on tactics, building construction, fire behavior, and many other aspects of the fireground. This is a unique position since other companies on the fireground are focused on their tactical assignments and do not have the time and "bird's eye" view of the fireground that the RIT has. Many firefighters and officers have never had the luxury of being able to witness a fire scene to understand a fire chief's point of view. By educating the team, they can then begin to understand the difficulty of reading the outside of a fire building in an effort to determine the progress on interior companies, collapse potential, fire spread, and much more.

Deployment in adverse weather conditions

Weather, such as extreme cold, deep snow, ice, extreme heat, humidity, and heavy rain, can all impede the ability of the RIT to deploy. The RIT must work with the incident or RIT commander in locating a sheltered staging area that would be close enough and effective enough to shelter the RIT from adverse weather conditions. Examples of shelters have been office or apartment building lobbies, residential garages, air conditioned or heated fire apparatus, and rehab buses. In many cases, RIT shelters have been left to the imagination of the team members.

Extreme cold

Depending upon the type of winter weather conditions such as cold, ice, sleet, heavy snow, and wind, it will be important for the RIT to be somewhat sheltered. The team and their equipment must be in the best condition possible if they are needed to deploy. Heavy, deep snow can also hamper the RIT, in that it can fatigue team members and cause them great difficulty just to transport their tools from their apparatus to the fire scene to stage.

Extreme heat

During routine staging of the RIT during a structure fire, the RIT can dress down from their turnout gear and SCBA. Once in the staging area, the RIT can set up their turnout gear in a ready position in the event they must deploy. With summer heat conditions, let common sense prevail. If the team were needed to deploy after standing in their turnout gear for ten minutes prior to the deployment, they would themselves, in this instance, fall victim to heat exhaustion.

Rapid Intervention Size-Up Operations

Each Mayday distress call is unique and is subject to the ever-present Murphy's Law, which simply states, "Anything that can go wrong, will go wrong." This is even more reason to remember to exercise caution in activating a RIT. When a Mayday distress call goes out or a PASS alarm activates, it should be remembered that any number of things can go wrong (e.g., a flashover or secondary collapse still can happen, crippling the RIT).

Fire service personnel need to recognize that rapid intervention operations are combatant positions as opposed to "administrative fireground support." Other than the deployment of a RIT for a rescue, the size-up operation is very aggressive, frequent, and it involves the entire team. This is when the RIT officer begins to establish the leadership role and brings the team together.

Initiating the RIT officer size-up procedure

1. The RIT officer or RIT Sector Officer (if assigned) should immediately report to the IC.
2. Establish a staging area for the RIT.
3. Obtain a briefing from the IC.

4. Inform the team of the staging area, the needed RIT tools, and the need for a walk-around size-up.

Fig. 2–3 The RIT Officer Performs a Complete Size–Up Around As Much of the Building As Possible (Hervas)

5. As part of the team retrieves the RIT tools from the apparatus, the RIT officer initiates the size-up around the building. To eliminate any accusations of members of the RIT freelancing, the need for comprehensive standard operating guidelines must be followed for size-up procedures. Additionally, it is imperative that direct notification be given to the IC that the RIT will be encircling the building.

RIT officer size-up

The walk-around size-up should be executed with the RIT officer and another team member or the RIT officer and the RIT sector officer. It is imperative to operate as a team of two for the following reasons:

1. Operating as a minimum team of two personnel with radios provides more safety.
2. "Two heads are better than one." In an effort to quickly observe and retain size-up information, two personnel is a definite advantage.
3. Preplanning possible rescue scenarios are easier, faster, and more thorough.
4. It provides an opportunity for mentoring. As the RIT officer brings a firefighter along on the size-up, it provides an opportunity for the firefighter to learn and experience more about fireground operations and how to pre-plan rescue operations.

The proactive RIT officer will then bring the team together and share the information concerning the size-up. Such information will involve:

- Immediate fire conditions – "winning or losing" the fight against the fire.
- Building construction type, occupancy, and dimensions.
- Elapsed fireground time.
- Effectiveness of present tactics.
- Building access (e.g., doors, windows, porches, fire escapes, and ladders).
- Special concerns (e.g., high security, fences, masonry deterioration and cracks, building exposures, electrical wires, etc.).
- The building preplan. (The RIT officer should attempt to locate a building preplan. This will be most important for large scale commercial, industrial, and high-rise buildings and can answer many questions that cannot be answered by a physical size-up, such as the interior lay out the building or a particular floor.)

Review size-up rescue scenarios

Once the size-up information has been collected and the team is assembled, it is important for the officer to put the information to work by preplanning possible rescue scenarios. For example, at a multiple alarm fire that was in the rear of a one-story grammar school measuring about 250 ft X 100 ft with masonry construction, the RIT team had been ordered to assemble, stage the tools, and begin the size-up. The fire was in a gym storage room where burning rubber mats were located next to the main electrical room. During the size-up, it was found that the building could not be encircled because of a heavy fence (unless the fence was to be cut), the roof was pre-cast concrete, and the electrical room still had live electricity. After approaching the IC, it was immediately recommended that a second RIT be positioned on the opposite side of the building. The team briefing focused on the difficulty of reaching the attack companies due to the live electricity, difficult ventilation situation, and the possibility of wide-area search operations.

Communications with the second RIT were then established, and another size-up was done to reevaluate the tactics and the structural conditions. Then, another team briefing took place involving updates and possible scenarios involving firefighter rescues. The first scenario reviewed involved a Mayday distress call for a missing/lost firefighter. Given long hallways and a gymnasium filled with smoke, the wide-area search was reviewed. Anchor points were located, use of secondary ropes was reviewed, placement of the thermal imaging camera was decided, communication techniques restated, and the need for additional support was preplanned. The second scenario reviewed a firefighter who could be entangled and low on SCBA air. Again, the need for a thermal imaging camera, search rope, and resources for additional SCBA air management had to be considered.

In many cases, by the time the second walk-around size-up takes place, the team has been briefed at least two times, the fire has been declared under control, and the RIT can stand down. Fortunately, the RIT was not deployed in this case, but far more important, was that they were combat-ready at all times.

The fact is that if the team is focused on their mission, ready with their tools, and mentally prepared to deal with almost any imaginable situation, it will result in the safest and most effective deployment possible. This does not mean that firefighters will not become injured or that the rescue will always be successful, but the likelihood of minimizing losses and affecting a rescue are at their greatest because the RIT is operating at an optimum 100%.

Check the building dimensions (width x depth x height)

Upon arrival, the RIT officer should assess the amount of area that must be dealt with in terms of the building dimensions. As the dimensions are calculated in the street, it is best to round off (e.g., 200 ft X 300 ft X 3 stories high) as opposed to trying to be exact by figuring 175 ft X 250 ft. In most cases, such exact dimensions are never correct, nor do they really make a difference in their application. The building dimensions will mainly help the RIT do the following:

Fig. 2-4 In Figuring the Building Dimensions for Fire Fighting and Rescue, "Round Off" (Hervas)

1. Calculate how much hose it would take to stretch if the RIT or support teams needed a protection line.
2. Calculate the amount and number of lengths of search rope needed and whether a wide-area search plan should be implemented.
3. Calculate the amount of SCBA air needed to execute various rescue scenarios. In some cases, it might be advisable to use 60-minute SCBA cylinders for the initial RIT.
4. Calculate the number of support RITs needed for various types of rescue scenarios due to the area and/or height of the building.
5. Calculate the need for additional RITs that might need to be staged in strategic points around and inside large buildings.

Using the dimensions of the building, the RIT officer can begin to review several potential firefighter rescue scenarios. As an example, the RIT officer's anticipation that the attack companies can commit to deep inside a large building to a "point of no return" will become a great concern. There is an increased chance that the RIT may be needed because the attack team(s) may consume too much of their SCBA air before turning around or that firefighters may also become disoriented and lost inside the large building. Additional concerns of the RIT will involve the chances that firefighters operating in a large building can fall victim to severe fire conditions, heat exhaustion, fatigue, or even heart attack.

Check the building occupancy

Building occupancies can be broken down into the following categories:

- Residential single-family
- Residential multiple-family
- Stores and office
- High-rise
- Warehouse and manufacturing
- Institutional and educational
- Public assembly

The RIT officer should immediately determine how the building occupancy would affect the fire fighting operations and any potential problems. If the occupancy is a hotel or apartment building, the RIT should anticipate the risks that search teams might be taking. Operating above the fire floor conducting a primary search without hoselines and from laddered windows is a standard procedure in many fire departments. The RIT will have to stay alert as to where the search teams are located, how many are searching, and listen constantly to the fireground radio communications for updates.

If the building occupancy is the warehouse and manufacturing type, the civilian count may be minimal, yet the building may harbor dangerous chemicals and various "mantraps." This will further require the RIT to constantly size-up and anticipate such dangers. In buildings of this kind, there are often confusing office layouts that are maze-like along with wide-open areas. The RIT may find some assistance with the building layout by checking a building pre-plan at the command post.

Check the building construction types

While sizing-up the building, it is extremely important to determine the building construction type(s):

Unprotected wooden frame. The walls, floors, and roof structure are wooden framing. There is no fire-resistive covering protecting the wood frame or sprinkler system.

Protected wooden frame. The walls, floors, and roof structure are wooden framing. The interior walls and ceiling surfaces of habitable spaces are protected by a fire-resistive covering, or protected by sprinklers.

Unprotected ordinary. The load-bearing walls are masonry. Columns, wood floors, and roof decks are exposed and unprotected from the effects of fire.

Fig. 2-5 Unprotected Wooden Frame Home (Hervas)

Fig. 2-6 Protected Wooden Frame Structure (Chief Steve Brown, Adams (MA) Fire Dept)

Fig. 2-7 Unprotected Ordinary Building (Ken Wood)

Fig. 2-8 Protected Ordinary Building (Kolomay)

Protected ordinary. The load-bearing wall is masonry. Columns are protected by a fire-resistive covering. The underside of all wood floors and decking is protected by a fire-resistive covering. A sprinkler system may also provide additional protection.

Unprotected non-combustible. A totally non-combustible building where the structural steel is exposed to the effects of a fire.

Protected non-combustible. A totally non-combustible building where structural steel is exposed. All vertical openings are protected by approved doors. The fire-resistant coverings of the steel are typically light (e.g., gypsum board, sprayed fire-resistive covering, and similar materials).

Fig. 2-9 Unprotected Non–Combustible Structure (Kolomay)

Fire-resistive. A totally non-combustible building where none of the structural steel is exposed and all of the vertical openings are protected with approved doors. The fire-resistant coverings of the steel are typically poured concrete, brick, or hollow concrete block.

Heavy timber. A typical mill-constructed building. The load-bearing walls or columns are masonry or heavy timber and all are exposed wood members, which have a minimum dimension of 2 in. If steel or iron columns are present, they are normally protected by a fire-resistant enclosure.

Fig. 2-10 Protected Non–Combustible Building, Such As the Steel Constructed 100–Story John Hancock Building (Sully Kolomay)

Check the building construction types as related to smoke and fire behavior

Two very important questions that the RIT must answer are:

1. How long will the fire (spread) tolerate the presence of firefighters inside?
2. How long will the construction (while attacked by fire) tolerate the presence of the firefighters inside?

For example, a fire-resistive compartmentalized apartment building can contain and withstand a great amount of a fire for an extended period. In this instance, the most important factor for the RIT to consider is the immense amount of heat and potential for burns that could victimize the attack and search companies. In contrast, an unprotected wood-frame building constructed of lightweight wood I-beams with composite board decking will only withstand moderate fire conditions for a very few minutes before the structural components disintegrate and collapse. In this instance, the most important factor to consider is the much greater potential for firefighters to fall through a floor or roof. It is imperative that the RIT understands and is experienced with building construction in relation to fire behavior. This will allow the RIT to better anticipate potential problems and react appropriately when needed. In order to understand how building construction relates to fire behavior, you must understand the behavior of the structural components in a fire.

Fig. 2-11 Fire–Resistive Building of Steel and Concrete Construction (Kolomay)

Fig. 2-12 Heavy–Timber Constructed Structure (Kolomay)

Fig. 2-13 Smoke Conditions Can Indicate Some of the Most Important Aspects of the Firefight and Potential Need for RIT Deployment (Hervas)

Fig. 2-14 Dimensional 2 x 10 in. Roof Rafters (Hervas)

Wood. Wood will add fuel to the fire load and even more so when exposed wood finishes are exposed in terms of increasing flashover times and temperatures. The greatest concern to the RIT will be wood and composite materials used for lightweight truss construction. In a combined effort, the Illinois Fire Service Institute and the Champaign (IL) Fire Department 5 conducted extensive tests on five different floor systems measuring 8 ft. x 9 ft., with tongue and groove wafer board decks. Each system had a live load of 31 psi distributed across the deck and was suspended over a Class B fire source and allowed to burn to failure. The failure times were recorded and compared as follows:

Wooden floor system	Sag time	Burn-through or failure
1. 2 in. x 10 in. dimensional floor joists	8 minutes	9 minutes
2. Wooden I-beams		4 minutes 40 seconds
3. Open web truss with wooden 2 in. x 4 in. cords and webbing with metal gussets	8 minutes	9 minutes (did carry the live load for over 15 minutes)
4. Open web truss with metal stamping on the outside for a weband wooden top and bottom chords	6 minutes	7 minutes 30 seconds
5. Open web truss, wooden top chord and steel pipe for webbing	6 minutes 50 seconds	9 minutes 45 seconds

1. 2-in. x 10-in. dimensional floor joists: Floor sagged at 8 minutes and failure occurred at 9 minutes.
2. Wooden I-beams: Failure occurred at 4 minutes and 40 seconds.
3. Open web truss with wooden 2 in. x 4 in. cords and webbing with metal gussets: Sagged at 8 minutes and failure occurred at 9 minutes. The system did carry the live load for over 15 minutes, but heavy fire conditions resulted from openings throughout the system allowing fire to travel freely.
4. Open web truss with metal stamping on the outside for a web and wooden top and bottom chords: Sagged at 6 minutes and failure occurred at 7 minutes and 30 seconds.
5. Open truss wooden, top chord and steel pipe for webbing: Sagged at 6 minutes and 50 seconds and failure occurred at 9 minutes and 45 seconds.

Fig. 2–15 Wooden I–beam Construction (Kolomay)

Fig. 2–16 2 x 4 in. Truss Beams with Metal Gusset Plates (Hervas)

If, after the initial minutes of the fire attack, the RIT officer could identify the lightweight construction, the fire load, and approximate burn time, the potential collapse can be calculated. Realizing that the average lightweight structural collapse time is within approximately 3 minutes, the risk of such collapse can be somewhat determined. It is additionally known that the average burn-through time of the floor decking is less than the joist failure time. A failure of either of these components could result in firefighters falling through floors. Armed with this knowledge, the RIT could be on its highest alert, thus increasing the likelihood of success.

Fig. 2–17 Metal–Pipe Webbing Beams Supporting Roof of a Future Restaurant (Kolomay)

Steel. Steel will not add fuel to the fire load. However, it will heat up, expand, twist, and fail, dropping the load it was designed to carry. In addition, as steel beams expand they push out walls, which causes structural collapse. Typically, an I-beam will expand approximately 9 in. at 1000° F, with resulting failure occurring at 1300°F.

By observing steel beams or steel-bar truss beams, comparing the fire conditions, watching the time, observing the effect of the hose streams, and watching the walls, the RIT officer can quickly discern how well the building will or will not stand. However, the experienced eye of a seasoned firefighter can still be fooled by a sudden roof collapse with steel beams.

During the winter of 1984, three Chicago firefighters were killed at the Vicstar Electronics outlet store when the roof collapsed during a fire. The 30 ft x 100 ft building, one and two stories in height, built of unprotected ordinary construction, was the target of an arsonist who used flammable liquids throughout the building to ignite the fire. The flat roof was very solid, supported by steel I-beam joists that spanned the width of the building. The roof ventilation operation was routine. However, entry into the building proved to be very difficult because of the high security systems. This delay allowed the fire to deteriorate and weaken the roof rafters in the cockloft. What was not known (and against building code) was the mounting of a large air conditioning unit on the roof. It was between the steel I-beam joists instead of on the joists themselves. If it had been installed according to code, the unit would have safely transferred its dead-load weight to the load-bearing walls. The combination of the weight of the snow, the firefighters and their equipment, the heavy fire, and the air conditioning unit caused the roof to collapse without warning.

Although there have been several guidelines offered to help determine when and how a steel-bar truss roof might collapse, those guidelines have been ineffectual. To the RIT officer, a steel-bar truss is unpredictable trouble when trying to evaluate the scene. Such roofs have collapsed without warning within five minutes of arrival of the first engine. As stated best by Mr. Francis Brannigan, "A truss is a truss is a truss. They are killers."

Concrete. Concrete will not add fuel to the fire load. Instead, it will spall, expose any re-enforcing steel, and eventually fail, dropping its intended load. It will also hold great amounts of heat and instantly convert water to steam. It is imperative that the presence of concrete in a fire be considered in all rescue operations.

Check the potential for collapse

RIT personnel need to be aware of areas of potential collapse during interior overhaul. Interior stairwells, exterior porches, and floors that have been severely damaged could collapse and, subsequently, weaken or collapse walls as well. If exterior streams were used to knock down heavy fire and interior overhaul is to be performed, the building must be surveyed from top to bottom. While water is draining from the building, aerial devices should be used to assess the structural conditions on the roof, the stairwells and floors, and the overall conditions of the exterior walls. During this assessment, only chief officers and personnel operating the aerial(s) are allowed in the collapse zones. At this time a RIT must be in position and very alert. **We must remember, buildings can be replaced but firefighters cannot.**

Check the placement of windows, doors, fire escapes, and porches

As the RIT officer completes the initial rapid "walk around" to size-up dimensions, occupancy, and construction type(s), points of entrance and exit must be noted as well. These various windows and doors are potential points for escape, rescue, and entrance for rescue operations.

Check the potential dangers of high security doors, barred windows, and building modifications

Such blockade dangers need to be addressed and sometimes dealt with immediately depending on the situation. If the fire is being attacked from the front (or burning side) of a downtown commercial store 35 ft x 60 ft and two stories in height, and the rear first floor access door is heavily secured, the RIT officer should inform the IC immediately. If a company is not available to open the door, the RIT should be ordered to open the door, then return to their staging area. The concerns in this case and others when a RIT is assigned to any fireground are:

- If the RIT is occupied and focused on a difficult forcible entry problem, the whole purpose of having a RIT is negated.
- Using the same example of forcible entry, the RIT can conceivably become fatigued and possibly use up their valuable SCBA air in the process.
- If the RIT is out of position and not paying attention to the changes in the fire conditions and the structure, they can just as easily become victims as much as the firefighters they have been deployed to save.

Based on experience and staffing, the RIT officer will have to make a judgment as to whether or not to attempt to force open a door, raise a ladder, or move a hoseline. If the task of opening a high security system on a door is too labor-intensive, then the RIT officer must immediately notify the IC.

Check the fireground tactics: offensive, defensive, defensive-to-offensive, or no attack

As soon as possible, the RIT officer should start to determine what type of tactics are being used by the IC.

Offensive attack: The RIT can anticipate that firefighters are operating inside the building and will have to use radio communication, progress reports, fireground time, and visual observation to determine any tactical progress. Depending on the standard operating procedures of the RIT and the type of building, the RIT officer and another member can enter the building to perform reconnaissance. Locating numerous points of entrance into stairwells, through offices, and at elevator locations in large commercial, industrial, and high-rise buildings is one example of why such important reconnaissance needs to be done

Defensive attack: Companies attempting to confine a fast-spreading fire will be concerned about hastily retreating to safer positions and securing a water supply. Upon arrival, the RIT will have to immediately investigate the status of the company or companies that arrived on-scene first. If an interior attack has been attempted and then the firefighters backed out of the building, did they definitely get out? Where are they now?

Many defensive attacks also require collapse zones. There are times when firefighters enter those collapse zones and are struck or buried by falling debris. Deployment of the RIT will be weighed against the likelihood of a secondary collapse. This is where the knowledge and experience of the RIT officer is needed for a split-second decision to determine if the team can survive the risks involved in a possible rescue.

During 1961 in Chicago, Illinois, nine firefighters were killed and 56 injured at the Hilker-Bletsch fire primarily due to a secondary collapse. The unprotected ordinary and heavy-timber constructed buildings occupied an entire city block with three different industrial occupancies. While changing to defensive tactics, a battalion chief and a firefighter were evacuating a two-story factory when a partial collapse of a six-story exposure building fell onto the building they were in and trapped them on the second floor. As they called for help from a barred security window, firefighters raised ladders and started to cut the bars with an oxyacetylene torch. Suddenly, secondary collapse debris fell onto the same building and killed the two trapped men and seven other rescuers while injuring and trapping many others. This was a situation where all of the rules could be thrown out the window, so to speak, and the combat-readiness of a RIT is needed most. In this case, some firefighters refused to get close to the building to assist in the rescue efforts, and others ran toward the building to assist as the building was falling apart around them. As can be seen in this example, secondary collapse is a primary enemy for the RIT. The need for education about secondary collapse is best evaluated by retired FDNY Deputy Chief Vincent Dunn when he states that it "should be mandatory training for all fire service personnel."

Another situation that could arise during a defensive attack is when the original fire is confined and an offensive attack is being carried out in an exposure building. In this case, it may well be that the direct attention of the RIT is needed at the secondary location instead of at the site of the original fire.

Defensive-to-offensive attack: This attack carries the greatest risk because the interior companies operate inside the building immediately after a "blitz," that is, after a master stream has been used to confine a fast-spreading fire. The building may be severely weakened in areas by the fire and loaded with water. The building construction and the occupancy type will help the RIT if this choice of attack is chosen.

Check the tactical command board

Use of the command-tactics operations board is another source of tactical information for the RIT officer. Does the observable information match the written command information? The tactical board can display great amounts of information rapidly (if used properly) that can help determine the number of attack lines, positions of companies, search operations, etc. Another source of information from the command post is the fireground accountability system. In anticipation of a catastrophic event such as collapse, the RIT can account for an estimated number of firefighters committed to the scene, their assignments, and what company they are assigned to.

Yet, another way for the RIT officer to obtain information is from the IC. Conversely, updated information or various concerns can be expressed to the IC from the RIT officer regularly. This relationship between the IC and RIT officer during an incident can become especially valuable if the RIT is activated. In addition to verbal communications, observing the general behavior of the IC in terms of calmness, confidence, confusion, anxiety, and many other types of behavior at an emergency scene can indicate to the RIT officer the true situation.

Check the fireground ladders and truck company operations

A critical question that is a constant concern of the RIT officer is "Are there enough truck operations going on at any given time?" There must be enough personnel and designated companies to provide the needed support for the engine companies operating in the interior of the building. Of course, the more aggressive the interior attack, the greater the need for ventilation. Correspondingly, the more ladders raised, the better the chance for a firefighter to be able to escape from a dangerous situation. If, for some reason, the interior companies are not receiving the needed truck support, the RIT officer should suggest to the IC the need to raise more ladders. If there is a shortage of personnel at that time, the RIT may have to raise the needed ladders. Realizing that the use of a RIT for fireground tasks is not encouraged, raising ladders is done as quickly as possible with as little energy as possible. Since this action benefits both the interior firefighters and the RIT, it is one fireground task that the RIT is encouraged to perform.

Check fireground time versus fireground progress

The time of alarm will have great significance to the RIT throughout the fire in relation to fire behavior, building construction, and firefighter endurance. Since the RIT usually arrives after the initial companies have committed to their fireground operations, the RIT officer will have to obtain the time of alarm by noting its dispatch time using CAD, requesting the time from the IC, or via radio to the dispatch center. Experience has shown, the more elapsed fireground time, the greater the chance of a problem occurring with a firefighter.

One naturally occurring event that can help determine the time that the initial companies have been committed to the fireground operations is the low-air alarms on the SCBA that the firefighters are using. These generally sound within approximately 20 minutes from when they were first activated (based on a 30-minute air cylinder). If these alarms sound after 20 minutes and the fire is still not under control, then the RIT should be extremely observant and ready for potential problems.

Check with the rehab sector

During an extended operation, check with the rehabilitation officer to assess the overall condition of the firefighters. The following questions need to be asked:

- What is the average recovery time?
- Is there a rehab rotation occurring?
- Is anybody requiring medical service for heat exhaustion or exposure to cold?

As firefighters physically deteriorate, the chance of injury increases. Although the RIT may not be able to control this situation, they must be aware and prepared to deal with it.

Check with the safety officer

The safety officer often shares the same concerns as the RIT officer. Therefore, it is important for the two officers to have frequent contact in an effort to share information and compare notes. The main difference between the safety officer and the RIT officer is that the safety officer will evaluate and correct safety concerns, whereas the RIT officer will not only evaluate safety concerns, but also prepare for various firefighter rescue scenarios in the event of a catastrophe. Because of the common safety concerns that both of these officers share, there has been some discussion that perhaps the RIT should report to the safety officer instead of the incident commander, thus adding another link to the chain-of-command. This is not advisable for the following reasons:

- The safety officer is generally mobile due to evaluating the structure and personnel and may not always be readily available.
- The safety officer may not always be a chief officer possessing the necessary command and control authority to activate a RIT.
- Placing the safety officer ahead of the IC has proven to only add another level of bureaucracy in the chain-of-command that is intolerable during a Mayday distress call. A direct line of communication must exist between the IC, the RIT, and its operations.

Check with emergency medical service (EMS) personnel

An emergency medical service (EMS) unit should be on the scene and designated for the welfare of the firefighters operating on the fireground. Initial response to a structure fire may call for the use of one or more EMS units due to civilian injuries from smoke inhalation, burns, or people jumping from windows. Just as quickly as civilians may be in need of EMS, firefighters may also be in need of the same services. In many cases, if a firefighter is in need of EMS, the RIT may also be activated even if it is just to assist an ambulatory injured member to walk around the building to the EMS. During extended fire fighting operations, EMS will be in even more demand due to the increased possibilities of firefighter injury because of the increased time spent on the fireground. Added to this is deteriorating structural situations, firefighter fatigue, and many other factors. If at all possible, an EMS unit should be on the scene and designated for the firefighters operating on the scene. The first place the RIT will go after they have been activated, deployed, and have recovered a missing or trapped firefighter is to the EMS unit.

EMS training has also played an essential role in firefighter rescue inside the building or at the area of rescue where the victim must be medically stabilized during an extended extrication or collapse recovery operation.

Relocate or add another RIT

In some cases, such as in a large factory with a fire in the rear, the incident commander may be located in the front of the building. It might be highly advisable to suggest that the RIT be relocated to the sector officer in the rear of the building to eliminate response time, distance, lack of tool availability, and fatigue. The option of an additional RIT team may become a necessity if the building has barriers around it (e.g., fences, walls, railroad spurs, etc.) thus prohibiting the RIT from being able to reach all sides. Given a sizable building, such as a shopping mall, industrial warehouse, or high-rise, the need to relocate or add another RIT will be equally necessary.

There have been several situations with large building fires where the RIT was not staged near the working area where the greatest concentration of fire fighting was being conducted and a Mayday distress call had gone out. In one case, the RIT could not accomplish their mission because of the time it took to get around the building, and the existing personnel had to struggle to rescue the fallen firefighter. In another case, the RIT not only experienced a delay in getting to the rear of the building due to the building's size, but also because of a breakdown in communications from the rear to the front after the roof collapsed with firefighters inside the building. Because of circumstances like these, it is essential that the RIT officer address the issues of where the staging of the RIT should be and whether additional RITs will be needed. These decisions should be based on:

- Size and complexity of the building (e.g., high-rise, shopping mall, factory)
- Ability to reach all entry points around the building
- Number of buildings involved
- Area(s) of heaviest concentrated fire fighting

Deploying and staging the RIT

Depending on the overall situation at the fire scene, the RIT stages in the most advantageous place possible. There is no ideal place, but there are some basic strategies that can be used according to the type of structure(s), the access point(s) available, and the information known about the structure.

High-rise buildings. If operating at a high-rise fire, the RIT could position near the interior IC one or two floors below the fire floor. Depending on the floor area (in some cases 20,000 square feet), RITs can be staged at specific stairwells closest to the fire fighting. Any additional RITs needed would stage several floors below the fire at a command and logistics sector.

As the RIT officer is staging, information about floor layouts, stairwell locations, window arrangements, elevator access, and any other special concerns pertinent to the high-rise can be obtained.

It will also be essential for the RIT to transport their tools and equipment upward (hopefully by elevator) to an assigned staging area. Again, it will be well worth taking the extra time to itemize the RIT tools before going up to the staging area. Not only would it be unrealistic to waste precious time to take the elevator back down 20 floors to get the missing Halligan bar, but by the time you realize that it is missing, the elevators might not be working anymore.

Shopping malls. At a shopping mall with multiple entry points and different levels of entry, the primary RIT should be positioned closest to the area of the fire fighting. In the "super" malls, the primary RIT can be positioned inside the mall near the interior sector commander as the attack teams suppress the fire. Again, the deployment time and distance between the RIT and the front line of the fight is an important concern.

As the RIT officer is staging, information about floor layouts, stairwell locations, window arrangements, elevator access, and any other special concerns pertinent to the interior mall can be obtained.

Additional RITs could be staged with other sector commanders inside a building of this size to cover different points of entry in a wide-area search operation.

Piers. Pier fires offer limited access into fire areas and equally limited escape and rescue opportunities. Piers can be very large and be the site of shopping malls, auditoriums for entertainment, carnivals and boardwalks, or industrial buildings containing heavy machinery. Logistically, piers generally allow entry only from the landside, thus limiting the attack positions of fire apparatus and personnel. A predominate danger when fighting a pier fire is having attack or search companies become separated and cut off by fire. The ability to affect a rescue is greatly delayed because aggressive hoselines must knock down the heavy fire so a rescue effort can proceed from the landside of the pier.

A logical place to stage a second RIT would be on a fireboat with a full complement of tools. During the efforts to fight the fire, if the fireboat is committed to using master streams, it can shut down and move the RIT into position from the most accessible waterside mooring closest to the Mayday distress call. The fireboat can supply additional hoselines, tools, communication, and become the apparatus of rescue for the RIT operation.

Factories and warehouses. Complications with factories and warehouses occur, not only because of their size in floor area and height, but also because of inaccessibility. Many such buildings have no windows and contain mantraps, high security systems, roll down doors, and maze-like interior configurations. To commit a RIT into a building of such floor area to search for a missing firefighter will require good radio communications, a building preplan, a wide-area search procedure, and adequate personnel to execute and support the search effort. If the rescue effort falls short in any of these areas, there is a high risk that more firefighters could be lost. If the RIT is committed deep within the building where there is cold smoke on the floor and poor (if any) visibility, the biggest enemy will be the limited amount of SCBA air and the rate at which it is consumed by the RIT(s). There is a high probability that the RIT may "over-commit" as they follow the victim's radio distress calls and/or an activated PASS alarm. Additional RITs entering from different access points can help reduce the occurrence of over-commitment and find the victim much more quickly.

Relocating and/or adding additional RITs will be a judgment call by the RIT officer in many cases. The need to move might be immediate or it might develop later in the incident. In either case, it can determine whether a RIT deployment to rescue a missing or trapped firefighter(s) is successful or not.

Rapid Intervention Checklist

If the RIT officer has been trained to stay alert and think ahead, the position of the RIT will serve as an invaluable life insurance policy for everyone operating at the scene of a fire and other types of emergency incidents. Naturally, as time marches on, memorized information and skills can deteriorate. For that reason, the use of a basic one-page checklist outlining the local RIT policy is recommended. The RIT checklist can be reduced in size, laminated, and affixed to the apparatus dashboard, visor, or any other accessible area of the cab for quick reference by the RIT officer.

RIT CHECKLIST

RIT SIZE-UP

Building Type and Dimensions (LxWxH)

Building Occupancy

Building Collapse Potential, Security Systems, and Access/Exit Points

FIREGROUND TACTICS

Offensive   Defensive   Defensive-to-Offensive   No Attack

Time of Alarm __________ First Entry __________ Elapse Time __________

Tactical Progress Report and Tactical Command Board or Sheet

Accountability

Laddering Around Building

RIT EQUIPMENT & TOOLS

Ordinary	**Fire Resistive/Non-Combustible**
Rescue Basket or Ladder	Rescue Basket or Ladder
Search Rope (150 ft)	Search Rope (150 ft)
Thermal Imaging Camera	Thermal Imaging Camera
SCBA Emergency Air	SCBA Emergency Air
Hand Tools (FE and Ordinary)	Hand Tools (FE and Concrete, Metal)
Power Saw(s) (Chain Saw/Wood Blade)	Power Saw (Steel Blade)
	Halligan Bar

OTHER RIT OPERATIONS

Emergency Medical Service

Safety Officer

Rehab Sector

Additional RIT or Relocate RIT

Review Possible RIT Rescue Scenarios

SPECIAL HAZARDS

____

____

____

____

Frequently Asked Questions (FAQs)

The formation of RITs is a real need that, if satisfied, can and will save lives. However, as mentioned in the introduction, the process for developing and instituting RITs has been a slow one. As authors of this book, we have conducted many seminars and instructed numerous courses throughout the United States. The following questions were gathered from the participants of those seminars and courses. It is hoped that these questions and their answers can facilitate the efforts of many fire departments to form their own well-trained RIT.

How can a RIT be staffed for response and operation?

- **Urban/Suburban:** Requires an additional response of another suppression unit from another battalion, district, division, or fire department.
- **Rural:** Requires an additional response of another suppression unit from another fire department or district. Due to the response distance, an immediate response is needed.

What is the recommended staffing for a RIT?

- Minimum requirement for staffing per NFPA 1500 6-5 is two firefighters.
- Depends on the expectations of the chief officers; do they expect a team of two to perform only size-up and tool staging, or a team of four to perform all RIT functions including recon and rescue operations?
- **Urban/Suburban:** Should be a minimum of four firefighters.
- **Rural:** Should be a minimum of four firefighters. Although the arriving suppression unit may not arrive with four, the team can assemble at the scene with available personnel.

Should the RIT be trained as a specialty team?

- No. Although certain large cities and metropolitan areas do have special heavy-rescue units that are staffed and trained in specialty areas (i.e., hazmat, confined space, collapse, etc.) this is a rarity. It is advised that urban, suburban, and rural fire departments/districts train all personnel equally, allowing all personnel to serve on a RIT.

When should a RIT be dispatched?

- **Urban/Suburban:** Can depend on the dispatch. For example, an activated fire alarm in an apartment building would not necessarily require a RIT response, but an activated alarm on the 18th floor in a high-rise might. The suggested average dispatch of a RIT would be upon the report of a confirmed fire.
- **Rural:** Due to lengthy travel and time responding from one district to aid another, it would be suggested to dispatch the RIT as early as possible. Upon the report of smoke inside the building or reported structure fire as opposed to a confirmed fire. Precious minutes could be lost!

Upon arrival, who should the RIT officer report to?

- **Urban/Suburban/Rural:** The IC. It is important that the RIT officer develop a relationship with the IC and answer only to the IC to eliminate confusion in the event of a catastrophic event.

Can the RIT be reassigned to fire suppression tactics?

- **Urban/Suburban:** Yes, but it is not desired. If the reassignment will make a significant impact in the safety and/or fire control of the incident, then it is highly suggested to reassign. If the impact of reassignment is negligible, attempt to avoid rotating units into rapid intervention operations.
- **Rural:** Yes, but it is not desired. Consider increasing initial responding fire suppression units to discourage the need to reassign and lose the RIT for an extended amount of time (travel time and distance).
- Personnel should not be assigned to rapid intervention operations for rehabilitation purposes.

When should rapid intervention operations be deactivated?

- **Urban/Suburban/Rural:** Should be based upon the IC's discretion. For a "quick attack" room and contents fire, the use of a RIT will be limited. If hazardous conditions exist during overhaul, a RIT would be advisable until those activities are completed.

While the RIT is staged in position, must the personnel remain fully dressed in turnout gear and SCBA?

- **Urban/Suburban/Rural:** This should be based upon the discretion of the RIT officer and/or IC. Whenever possible, the highest level of readiness should exist, but during extreme weather conditions or extended operations allowing "dress down" should be considered. It is counter-productive to expect personnel to remain fully dressed with SCBA in 90-degree weather and still perform RIT activities and execute a rescue operation.

What type of apparatus should respond and where should the RIT tools come from?

- **Urban/Suburban:** If possible, a heavy rescue unit or ladder company is suggested. Due to the amount of resources typically available, most equipment and tools can come from the arriving RIT apparatus.
- **Rural:** Although a heavy-rescue unit or ladder company is also suggested, limited resources may not allow these apparatus. Basic tools may have to be equipped on apparatus that does not normally carry such tools to accommodate RIT operations. Specialized equipment may have to be pooled together at the scene from other apparatus.

How should standard operation procedures for rapid intervention operations be written?

- **Urban/Suburban/Rural:** Response procedures should be included in dispatch procedures and rapid intervention operations should be kept to one or two pages at most. A one-page laminated RIT check sheet should be kept on the dashboard for quick reference and reminder.

3

Rapid Intervention Deployment Operations

The Recon Team

During the Vietnam War, the 101st Airborne Screaming Eagles were responsible for reconnaissance missions based on preconceived information. They gathered information about enemy movement, ammunition supplies, and communication bunkers. This information was used to help plan military strikes and in some cases rescue missions. In the context of fire fighting, the RIT reconnaissance (recon) team consists of the officer and several firefighters who carefully advance and search toward the assumed location of the missing, lost, or trapped firefighter. One of the first steps in gaining information is to ask the following simple and direct questions:

1. How many firefighter victims are missing?
2. What is the name(s) and assignment(s) of the victim(s)?
3. Where might the victim(s) be located?
4. Can the victim(s) be contacted via radio or verbal communication?

The New York City Fire Department has implemented the acronym LUNAR to help memorize the components of such information during a Mayday. The letters stand for:

L Location of missing, lost, or trapped firefighter

U Unit (company or department) firefighter is assigned to

N Name of downed firefighter

A Assignment of that firefighter (interior search, roof, etc.)

R Radio-equipped for communication or radio assisted feedback

Finding out such information before committing will expedite the rescue and help reduce the risk. However, reality and "Murphy's Law" may not give the RIT many of the needed answers. First sending in the minimum number of RIT members to perform a recon accomplishes a number of things. The knowledge gained can prevent committing additional personnel and tools to a wild goose chase or a dead end. This saves valuable time, energy, and SCBA air that the RIT cannot afford to waste. While reconnaissance is being conducted, the RIT recon will report back on:

1. Structural conditions and the ability to effectively and safely proceed into the building
    - The need to bridge floors or shore walls and ceilings will have to be determined in some cases.
    - A catastrophic collapse will require stabilization of surrounding structural members (e.g., free standing walls, lean-to floor sections, etc.) before a RIT recon team could ever consider entering the building to determine the whereabouts of a victim.

Fig. 3–1 The RIT Officer Conducting a Building Reconnaissance (Hervas)

2. Fire conditions that require the need of an engine company and additional ventilation
3. Other points of entry such as windows for using ground ladders, aerial ladder, tower ladder, or any other type of elevated platform
4. Any signs of life from yelling or banging by the victim
5. A PASS alarm, low SCBA air alarm, or radio-assisted feedback

Once the reconnaissance is completed, it will be the determination of the RIT officer as to whether additional team members should enter the structure or the RIT recon should exit and regroup for a briefing, to re-supply their SCBA air, and to gather tools and shoring.

The RIT Sector Officer and the RIT Rescue Sector

The position of a RIT sector officer is in addition to that of the RIT officer. However, the RIT sector officer is greatly needed when the RIT deploy for a Mayday distress call. Though most fire departments may not have the resources to assign personnel to a RIT, much less a chief officer to a RIT sector officer position, it has been proven to be a most valuable role during a rescue operation. Consequently, larger cities, which typically have larger and more complex buildings, have implemented the RIT sector officer position. For example, the Chicago Fire Department has assigned a battalion chief to be the RIT sector officer with their RIT on confirmed structure fires. The RIT sector officer duties include assisting in communication with the IC, setting up a staging area, conducting a size-up with the RIT officer, and reviewing rescue scenarios.

Fig. 3–1 RIT Sector Officer and RIT Officer Conducting a Size-Up at the Entry Point Before the RIT Deploys into the Building to Search for a Firefighter in Distress (Hervas)

A desirable candidate for the position of the RIT sector officer is one who has already been a ranking chief officer and is capable of overseeing and directing the RIT from arrival to decommitment from the scene. The main mission of the RIT sector officer is to provide the RIT with all of the support that it will need to accomplish its rescue. As simple as that may sound, it is not. The main reason why the RIT sector officer has been added to the RIT operation is because once the RIT officer enters the building, they become very focused in determining the risks as they crawl forward, account for the team, and accomplish the search and rescue. The intense concentration that the RIT officer must devote to these duties has made communication with, control of, and accountability for the rest of the team almost impossible. The scope of trying to communicate via radio, determining the surrounding fire conditions, maintaining accountability, contending with the urgency of the incident, and figuring out a search and rescue plan can, simply put, make even a basic firefighter rescue too much for the RIT officer to handle. **Careful examination in training has shown that a RIT officer operating without a RIT sector officer can easily lose a member of the team, lose radio contact, or not receive the support of additional equipment and/or help when needed. The duties of the RIT sector officer are:**

- Be radio equipped, have a large hand light, and be fully protected with turnout gear, PASS, and SCBA.
- Assume the command role of locating and rescuing the missing or trapped firefighter(s), thus allowing the incident commander to maintain control of the rest of the building and firefighting efforts.
- Control and account for all personnel at the point of entry whether entering or leaving the building.
- Stage and account for any additional RITs and special equipment (e.g., hydraulic tools) to support the initial RIT.
- Alleviate the RIT officer from having to use the radio to communicate if possible. The RIT sector officer will attempt to maintain face-to-face communication with the RIT officer as much as possible and relay information and needs to the incident commander by radio or face-to-face.
- Update and advise the RIT officer of deteriorating fire and/or structural conditions and the elapsed time of the search or rescue. It must be mentioned again that it is critical that the rescue be accomplished within the time limit of the SCBA that the RIT is using. If the RIT must be pulled out or relieved by another RIT, high levels of confusion and frustration can occur.

- Direct additional RITs into the rescue area and escort other teams out of the building.
- Control and account for original fire companies not associated with the RIT assisting in the search and rescue.
- The RIT sector officer works a position that is mobile between the entry point of the RIT and the rescue site inside the building. Extensive training has revealed the great need for the RIT sector officer to perform a brief interior reconnaissance to experience and personally view the interior conditions that the RIT(s) is experiencing. This experience has greatly benefited the RIT sector officers' decision-making from the exterior as well as the ability in determining the risks versus the benefits. If the RIT sector officer decides to remain in an interior position with the RIT, then that decision should be based on such incidents involving:

1. Difficulty in communicating with the RIT(s)
2. Need for coordination of multiple RITs for multiple victims
3. Extensive rescue operation involving disentanglement and/or extrication
4. Accountability of wide-area search operations
5. Difficult victim medical stabilizing, packaging, and removal

Many of these duties were once placed on the shoulders of the RIT officer. It is important to note that if the RIT sector officer must commit to the interior, another RIT sector officer assumes the exterior position immediately. However, it cannot be forgotten that the RIT officer may also be confronted with efforts to fight the fire, falling debris, and a limited ability to hear and see. Hence, the need for the RIT sector officer.

RIT Checklist

Designed to be used by the RIT Officer and/or RIT Sector Officer, the first page of the RIT Checklist on the following two pages is a worksheet to identify RIT personnel, confirm RIT positions and equipment, and evaluate the firefighting strategy and tactics. Basically, the first page of the RIT Checklist is a preparation before a possible "Mayday" is called. The second page is in the event a "Mayday" is called. The six-point Mayday Respond Procedure will assure that the proper equipment and personnel is responding, and the tracking chart will track the RIT accountability, entry point, and time of operation.

RIT Sector Officer Checklist

1. Incident Commander: ______________________________
2. RIT Sector Officer: ______________________________
3. RIT Suppression Co: ______________________________
4. RIT EMS Co: ______________________________
5. Collect Accountability Tags (Total RIT personnel): ______________
6. Time of Initial Alarm: ________ Box Alarm Time: ______________

- ❑ Size-Up Building and Fireground with RIT Company Officer
- ❑ RIT Staging Area Evaluation
- ❑ RIT Tools and RIT Personnel Positions Review
  - ❑ Position #1, Co. Officer, T.I.C./Radio/Halligan
  - ❑ Position #2, Rope Bag/150 ft./FE Tools
  - ❑ Position #3, RIT SCBA Air
  - ❑ Position #4, Tools
  - ❑ Position #5, Tools/Power or Hydraulic/Radio
- ❑ Special Hazards (e.g., security gates, glass blocks, power lines)
- ❑ Alternate RIT Rescue Preplans

- Identify Sectors
- Building LxWxH
- Construction Type
- Doors and Windows
- Fire Area(s)
- IC Position
- RIT Staging

Sector

Sector

Sector

Sector

Mayday Response Procedure

1. Fireground companies switch to alternate radio frequency
2. Escalation to the next alarm (e.g., Box to second alarm)
3. Additional EMS response
4. Collapse Rescue Unit
5. One (1) Additional Heavy-Rescue Squad
6. Back-Up RIT to RIT Sector (1-Truck and 1-Eng)

RIT Tracking Chart

RIT TEAMS	DEPTS/COMPANIES	ENTRY POINT/SECTOR	ENTRY TIME	5 MIN	10 MIN	15 MIN	20 MIN WARNING
RIT SECTOR OFFICER							
RIT #1							
RIT #2							
RIT #3							
RIT #4							
RIT #5							
RIT #6							
RIT SECTOR OFFICER							
RIT SECTOR OFFICER							

The RIT rescue sector officer

In a larger or more complex RIT operation, it is also a necessity to have an additional chief officer who oversees the rescue sector and is stationed at the command post. The RIT sector officer, while inside, can only alleviate the incident commander from the fireground confusion and communications problems for a brief amount of time. If the victim(s) are not removed quickly, the IC will become overwhelmed. The RIT rescue sector officer will then assume command of all rapid intervention operations, the RIT(s), and the RIT sector officer. This will allow the incident commander a better opportunity to reposition interior companies to protect the RIT search and rescue operations or to evacuate the building and conduct a personnel accountability report or role call.

The addition of the RIT rescue sector officer to the command post can also help an IC who may have been stunned by whatever catastrophe caused the need for a Mayday distress call.

The RIT officer and using the size-up information

The RIT officer should be aware of the building's conditions and the behavior of the fire until they are released from the scene. This will require the officer to physically size-up a building regularly. The experienced RIT officer knows that in addition to the basic observation of the fireground, he or she may have a "gut feeling" that occurs when something is wrong. These feelings or senses (sometimes referred to as the "sixth sense" or intuition) are not to be disregarded. As we know, a fire building presents itself as a living and breathing beast. Changes in heat levels, smoke color and movement, and fire behavior will indicate to the officer if the current tactics are working or if control of the fire is being compromised. The radio transmissions (or lack of), pace of the working firefighters, the smell of the smoke, and sounds of breaking glass will all add up quickly in the mind of the experienced officer and immediately produce a "gut feeling" that something is not right. The officer should immediately determine what is wrong and take whatever actions are needed to rectify the situation.

Fig. 3-3 The RIT Sector Officer Will Be Responsible for Maintaining Constant Communications with the RIT Operations, Providing Resources, Support, and Assist in the Many Decisions That Will Have to Be Made

Even after the fire has been controlled, the RIT officer must be aware that many firefighters have fallen victim to injury and death during overhaul due to building collapse, injury from tools, falls, heart attacks, and many other reasons. Although the RIT officer may be on the opposite side of the building evaluating the situation in that area, they should still be reachable by radio in the event the team is needed. The officer must be able to respond immediately to the incident commander at all times.

Fig. 3–4 A Structure Fire with Many Hidden Signs of Danger (The Waldbaums Supermarket, which occurred on August 2, 1975, claimed the lives of FDNY Lieutenant Cutillo, and firefighters Hastings, Bouton, McManus, Rice, and O'Conner when the roof collapsed during ventilation operations.) (Harvey Eisner, *Firehouse Magazine*)

Freelancing

Freelancing can be best defined as "being unaccountable in relation to personnel location, fireground duties, and communication." In many respects, RITs can be accused of freelancing due to the amount of RIT tasks and fireground movement that must be done before there is ever a need for a RIT. In other words, they are very mobile. Because of this mobility, a simple, practical, yet comprehensive standard operating procedure is needed to deter RIT team members from unintentionally freelancing or of being unjustly accused of it. Therefore, a quality training system becomes a must if all RIT rescue procedures are to be correctly followed by all members. Each RIT should develop a written operating procedure manual and include in it the definition of what freelancing is and is not. This helps to clear away any confusion that may develop as the result of the movement of the individual team members.

An example of when such confusion could occur is when the RIT is gathering and staging tools while the RIT officer performs a size-up around the fire building. To the untrained eye, it would appear that the RIT officer was acting alone and, thus, freelancing around the building. However, such reconnaissance around the building by the RIT officer is standard procedure. It is the only way for the officer to assess the conditions and report back to the other team members to inform them of any changes or further preparations they may need to make. At the same time, the RIT officer is careful to maintain radio contact with the IC, thus further reducing any misconceptions as to their activities.

Firefighter and RIT Accountability

Just as a financial accountant has to maintain checks and balances, personnel at the scene of a fire must also have checks and balances when dealing with emergency scene accountability. Simply put, as a firefighter or officer, it is just as important that you know your location as it is for you to know the location of your company. During normal firefighting operations, suppression companies will divide and conquer in some cases in an effort to confine fire, pull ceiling, ventilate, and search. However, such division is not recommended for the RIT. The following are several suggested interior accountability methods that can be used.

The RIT officer is "first in and last out"

A proven technique for accountability and control during RIT operations is that the RIT officer must lead the team into the incident. As mentioned in chapter 1, the "Presence of Command" must be visible and provide the needed leadership. Included in that leadership is the accountability of the RIT personnel. Understanding that accountability is a primary responsibility for the officer; the firefighters must also attempt to "check in" with the officer and stay with their assigned partner at all times.

The RIT must operate as a team

Actual incidents, case studies, and training scenarios have proven the need for a RIT to stay together while operating in the rescue mode. As defined earlier, a RIT should have a minimum of four firefighters. The officer and three firefighters will be involved in situations that are uncontrolled and unstable, as well as dealing with fire and/or collapse while attempting firefighter rescues. In many cases, the rapid intervention rescue efforts have proven to be very different from normal search and rescue methods for the following reasons:

- During the initial RIT reconnaissance inside the building, the RIT officer will have to set up a search rope, use a thermal imaging camera, communicate via radio, monitor conditions, and coordinate a search operation requiring a minimum of four personnel.
- Certain rescues will require firefighter drags and lifts, use of rope, disentanglement, extrication, SCBA replenishment, and other similar efforts. In almost all cases, it has been found that a minimum of four personnel are needed to rescue one victim due to equipment needs, quick stabilization and shoring, and the power to pull, lift, and drag.

Fig. 3–5 A Staged RIT with Tools (Kolomay)

- Air consumption is a concern since everyone on the RIT will be breathing somewhat differently depending on their workload, anxiety level, level of fitness, age, and interior temperatures. The RIT working as a total team can sometimes reduce the workload and at the same time expedite the rescue. It has been found in training exercises simulating realistic conditions that if a rapid intervention operation does not achieve its set goal of rescuing the missing or trapped firefighters within the limits of their first SCBA cylinder, the operation becomes endangered and the RIT also becomes endangered.

Assign RIT responsibility and tool assignments

Pre-designated responsibilities and tool assignments for the RIT is a very proactive procedure for any fire department to use. Similar to the responsibilities and tool assignments given to riding positions on truck companies, RIT tool assignments and responsibilities could be outlined as follows for a minimum size RIT of four personnel.

Position	Assignment	Tools
Position #1	Officer	Radio/Hand Light/Thermal Imaging Camera/optional search rope bag
Position #2	Rope Bag	Radio/Hand Light/Search Rope Bag
Position #3	SCBA	Radio/Hand Tools/Hand Light/Power Saw Personal Rope Bag/RIT Emergency SCBA Air (SCBA or RIT Pack)
Position #4	Tools (Entry)	Radio/Hand Tools/Hand Light
Position #5	Entry	Radio/Hand Tools/Hand Light/Power or Hydraulic Tool

(Optional position if staffing is available)

- Position #1, Co. Officer, T.I.C./Radio/Halligan
- Position #2, Rope Bag/150 ft./FE Tools
- Position #3, RIT SCBA Air
- Position #4, Tools
- Position #5, Tools/Power or Hydraulic/Radio

The RIT operations during deteriorating fire conditions

Most important is the ability of the RIT officer to maintain accountability of the team when they might become separated under deteriorating conditions. The RIT is an operational team that is supposed to resolve the problem, not contribute to it by losing more firefighters. If the RIT officer, or any company officer, loses accountability of their firefighters, they will be unable to finish their assigned task until those firefighters are found. This becomes especially critical when the assignment that must be abandoned is the RIT's search for a missing firefighter because the team has lost one of its own members.

Once the RIT has been activated based on a Mayday distress call, the incident commander should immediately review the location of all companies and personnel on the fireground. This can only be accomplished if an accountability system has been established by the department and is enforced at every incident. This accountability system should include a method of tracking the location and assignment of the RIT itself. If secondary rescuers or RIT members become unaccounted for during a Mayday distress call that was issued for other missing firefighters, several very important changes may take place, such as:

- The RIT and/or the officer of the originally lost firefighters will have to contact an already overwhelmed IC to communicate the need to locate additional missing rescuers.
- The IC will have to assign or reassign secondary rescue personnel to locate the missing rescuers. If the RIT is missing any of its members, then the rescue mission to locate the missing firefighters from the original Mayday is generally stopped.
- If the missing secondary rescuers or RIT members are not found quickly, the incident may quickly snowball into a disaster because the original missing firefighters have still not been found and more companies may become desperate to push into a deteriorating fire scene to attempt rescues on their own.

As mentioned before, we "never say never and never say always" because there may be situations that require the RIT personnel to separate briefly. Separation of the RIT may be necessary to collect and transport tools, raise ladders, size-up exterior operations, or conduct interior reconnaissance. Also, some rescues involving collapse situations may demand that some RIT members operate from a vantage point above or below the victim, thus separating the team. However, as a general rule it is recommended that the RIT not be separated because it threatens the success of the rescue effort, reduces accountability, and increases the overall risks to the RIT members.

Rapid Intervention Radio Communications

During post-incident critiques the most frequently criticized item is radio communications. Radio communications seem to falter and fail due to:

- Delivery of incorrect dispatch information
- Radio disruptions due to hardware problems including static, unclear transmissions, or reception problems
- Misuse of the radio by firefighters (e.g., too much needless radio chatter, yelling through the SCBA facepiece, poor choice of terminology or language, etc.)
- Radios not listened to because they are set down somewhere, lost, or put in a pocket. This usually results from the absence of a remote microphone on the radio or the absence of a department policy that addresses carrying and operating the portable radio

Fig. 3–6 Portable Radio Usage with a Remote Microphone (Hervas)

The importance of using a radio must be conveyed to every firefighter. Radio communications are critical. Many times successful transmissions have saved the lives of firefighters while failed transmissions have been responsible for the deaths of firefighters.

RIT radio communications identity

For rapid intervention operations, proper radio identity is crucial. A recommended method for radio identity for larger fire departments with company numbers:

"Squad #1 RIT" or "Ladder #142 RIT" or "Engine #98 RIT"

For areas that work with mutual aid and are acknowledged by the city, town, or district:

"Lake Zurich RIT"

This method of identification has proven to reduce confusion and has increased accountability. Although there might only be one RIT initially, it should still be identified with its company number or town (e.g., "Squad #1 RIT" or "Lake Zurich RIT"). If the team is deployed for a Mayday distress call, a backup RIT will be activated immediately. The backup RIT might come from staging, or it might be pulled from the fireground. In any case, it too should be identified in a similar manner. If the situation demands multiple RITs, each team should be identified as a "RIT," and the incident chief or sector chief should know every person who is assigned to each particular team.

Alternate radio frequencies. During a Mayday distress call, it will be imperative that all radio communication cease other than that between the firefighter(s) in trouble and the IC. It would be advisable to have all other fireground radio traffic switch to a different fireground frequency for the following reasons:

- To allow the original frequency to be open for the distressed firefighter(s)
- The original frequency to be used as the RIT frequency dedicated to the search and rescue operations
- The new fireground frequency to be used for fireground suppression operations

It is important to note that if firefighters have to switch frequencies they should all know what the second frequency is named (e.g., F2 or Fireground 2, etc.). In addition, the actual switching of radio frequencies should be made as simple as possible so that the firefighter can change to the new frequency with turnout gear on and without having to look at the radio. One way to accomplish this on a radio with a turn dial is to have the second frequency on the opposite end of the primary frequency. In other words, if the primary radio frequency is Channel #1 and there are 10 channels, then the secondary fireground channel should be Channel #10. This would allow the firefighter to simply locate the radio dial and turn it until it cannot be turned any more. Realizing that there are many other radio systems available throughout the fire service, we have found that "simple" is best no matter what the system.

Rapid Intervention Tool Selection

Responding rapid intervention apparatus

The fire service throughout the world varies in the type of services provided. Staffing options of volunteer, paid-on-call, combination paid-on-call/career, career, military, industrial fire brigades, and private companies exist. In addition to the form of the staffing, varying levels of service are provided by fire departments to address the social, geographic, structural, and physical features of their protected jurisdictions.

- Urban cities with factories, high rises, decayed areas, subways, and apartment buildings will typically have career staffing with a variety of up-to-date apparatus and equipment positioned throughout the city. Many urban fire departments are structured with engine and truck companies.
- Rural fire protection districts typically do not have the fire workload or financial resources to provide a fire service with career personnel or the newest of equipment. They usually will provide a dedicated volunteer or paid-on-call service having to work with fewer apparatus and minimum staffing. Aside from having problems getting firefighters to a fire scene, many rural departments are without fire hydrants and have long response distances.
- Industrial and airport fire protection requires very specialized training and apparatus to address chemical, jet fuel, explosives, and other special fires and accidents.
- Mountainous areas require apparatus, equipment, and training to address the transport of water into hard-to-reach areas up steep inclines and through snow.

Taking these many fire service variables into consideration, several recommendations can be made concerning the type of apparatus that a RIT should use in responding.

Aerial apparatus. Straight-frame aerial, tower ladder, ladder tower, Snorkel®, and other aerial apparatus are most desired for the following reasons:

1. The aerial apparatus will generally carry most of the tools that are needed for RIT operations (e.g., hand tools, heavy-rescue hydraulic tools, Stokes basket, rope rescue equipment, cribbing, power saws, etc.).
2. The aerial apparatus will generally respond with personnel trained in all of the equipment to be used and in search and rescue operations.
3. The aerial apparatus will have many of the needed ladders and the aerial device if needed.

Fig. 3–7 Aerial Apparatus (Hervas)

Fig. 3-8 Heavy-Rescue Apparatus Equipped for Firefighting, Extrication, Rope Rescue, Confined-Space, Trench, Hazardous Materials, and Collapse Rescue (Daniel P. Alfonso)

Heavy-rescue apparatus. Compared to an aerial apparatus, a heavy-rescue apparatus generally has more specialized equipment (e.g., torches and pneumatic lifting bags) but lacks a complement of ground ladders and the aerial devices.

Combination engine/rescue apparatus. Compared to an aerial and heavy-rescue apparatus, there is a lack of ladders, aerial devices, and some specialty tools. In most cases, the combination engine/rescue apparatus will be able to supply all of the basic tools for the RIT and has the advantage of being able to stretch hose if needed.

Engine apparatus. The engine apparatus is often used as a RIT by many departments because they are more readily available. It must be pointed out, however, that this apparatus is not nearly as efficient for the RIT as the aerial or heavy-rescue apparatus due to a lack of forcible entry and rescue tools, a deficiency of ladders, and the temptation for the IC to use the engine apparatus to provide a firefighting line rather than leaving the engine dedicated to RIT responsibilities.

Ambulance apparatus. This is a very restrictive option to serve as the responding RIT apparatus because of the lower number of personnel and lack of tools that are typically carried. Another problem on the fire scene is that ambulance personnel are often used to perform emergency medical services for civilians or firefighters, which may take away from the RIT's duties. There are fire departments that augment the RIT with an ambulance crew, but it is not recommended to use the ambulance exclusively as a RIT.

It is understood that every fire department, district, and brigade has different restrictions in staffing and equipment and that it may not be possible to always have an aerial or heavy-rescue apparatus available. At the same time, the demands of rescuing a firefighter(s) in a burning structure are often similar despite the differences within the fire service. We must try our best to reach for the most effective way to provide the right equipment and staffing even if it means reaching outside of your borders for help.

Rapid intervention tools

The amount of equipment deployed by the RIT should be determined by:

1. The size of the incident
2. The type and complexity of the incident
3. The level of risk in and around the incident at any given time

The key to deploying the proper RIT equipment is to combine the size-up information with the RIT officer's fireground experience. At first, the collection of RIT tools will conform generally to the type of structure involved. An unprotected wood-frame building would require axes, chain saws, wood cutting circular saws, and pike poles. A high-rise office building would have greater demands for longer search ropes, sledgehammers, bars, hydraulic tools, and emergency SCBA air supply.

Specific checklists for each of the most common structures provide the initial equipment list for deployment for the reconnaissance, primary search, and rapid rescue operations for a RIT made up of four personnel.

Additional RIT tools

If fire scene conditions are getting worse, the RIT officer should decide what additional tools need to be obtained. The following are tools normally used on a fire scene.

Hoseline. The RIT officer or sector officer can obtain a hoseline and additional personnel to help move it from another engine company to cover a possible deployment into the building. The additional personnel are needed because a RIT of even four members cannot feasibly stretch and move a charged hoseline. If staffing is low, the RIT officer may have to preplan what engine to stretch a line from and who will help to stretch it if it is needed. If additional engine companies are available, then a specific engine company should be assigned to the RIT.

Additional search rope. In a large-area building such as a warehouse, school, or high rise, consider the need for wide-area search operations. As the fire worsens, chances are good that firefighters have

Fig. 3–9 Hydraulic Spreading Tools and Cutting Tools of Various Sizes and Strengths

Fig. 3–10 Oxygen and Electrically Fed Cutting Rod Torch (Courtesy of Arcair ®, St. Louis, MO)

progressed deeper into the building and also have less SCBA air to use. This is when the RIT officer needs to preplan and collect additional tools such as more search rope. Not only will the initial RIT need rope, but other RITs may have to be deployed in an attempt to enter from other doors using whatever equipment the initial RIT has staged. Some fire departments have set up one large rope bag containing one main search rope bag (150 to 200 ft. long) and four to six additional bags of wide-area search rope (50 ft. each). This one large bag is something that could be transported immediately to the search entry point and deployed.

Hydraulic tools. There are a number of new hand-held hydraulic spreading, cutting, and lifting tools available that are light, mobile, and inexpensive. These tools can be staged with the RIT as conditions warrant. The RIT officer can usually locate any larger hydraulic tools that may be needed on other apparatus at the scene.

Specialized tools. Preplanning specialized tools such as torches, cribbing, collapse equipment, and technical rescue systems will be important according to the type of incident. Incidents where scaffolding, ladders, catwalks, and machinery are involved may be examples of when the RIT officer would preplan the location of such specialized tools, equipment, and trained personnel.

Transport of RIT tools

Since the RIT responds after the initial companies, the RIT apparatus may have to be parked around a corner or down the street. Once the RIT officer is directed to a staging area, the tools can be transported one of these three ways:

1. Hand-carried
2. Ladder tray
3. Rescue basket

Hand-carried. Individual firefighters carry tools to the staging area. The disadvantage of this method is that a limited number of tools can be brought to the staging area on the first trip. It is important that each firefighter on the RIT carries the maximum amount of tools in the safest manner. In many cases during normal operations, firefighters will not carry more than one tool into the building. However, they should carry at least two tools that will complement each other (e.g., axe and pike pole, Halligan bar and flathead axe, Halligan bar and plaster hook, etc.). In many areas where staffing is a critical issue, it is even more important that firefighters take the correct tools and the correct number of tools to do the job. There is no exception to this with rapid intervention operations.

Fig.3–11 Ladder Tray (Kolomay)

Ladder tray. A 24-ft. or 35-ft. extension ladder can be used as a tray to carry tools to the staging area. Many of the tools needed can be placed and secured onto the ladder. The advantage of this method is that more tools can be carried on the first trip to the staging area and the RIT then has a ladder added to their inventory. The difficulty with this method of transport is securing the tools to the ladder well enough so they are not falling off the ladder while being moved. With some training and experimentation, tools such as pike poles, power saws, and bagged search ropes can be fastened very securely onto the ladder. Generally speaking, sledge hammers, axes, and other selected hand tools still have to be hand-carried.

Fig. 3–12 Rescue Basket Used to Carry RIT Tools to Staging Area (Kolomay)

Rescue basket. The advantage of using a rescue basket is that all the initial RIT tools can be placed into the basket and transported to the staging area. Experience has shown that this is the most preferred method of transporting the tools. Other advantages to this method are that the tools will not fall while being carried and a full complement of tools can be carried quickly in one trip. When there are four firefighters on the RIT, two members can carry the rescue basket, and the other two can carry a ladder to the staging area.

Fig. 3-13 Rescue Basket Containing RIT Tools in Staging Area (Hervas)

RIT tool staging

The RIT tools that are staged should come from the RIT apparatus as opposed to collecting them from other apparatus. If the tools are taken randomly from other apparatus, there will be inventory problems, familiarity problems, and possession problems. Most importantly, you can imagine the situation that could develop if a firefighter goes looking for a specific tool on their apparatus and finds it missing or laying on the ground in the RIT staging area.

As mentioned earlier, the rescue basket is the most preferred method of transporting the RIT tools. Additionally, it is also the most preferred method of staging them as well. The advantages of using a rescue basket for staging are:

1. The tools can be kept packaged together so they are not taken by other on-scene firefighters. This reduces the opportunity for a firefighter working hard inside a building who has forgotten an axe to step outside, see the RIT axe laying on the ground, and take it.
2. The tools can be arranged inside of the rescue basket in an organized manner. For example, the flathead axe can be paired with the Halligan bar, the search rope bags can be grouped together, and the hydraulic door opener can be matched with the emergency air supply.
3. A rescue basket itself is a valuable piece of rescue equipment for rapid intervention victim rescue.
4. When a rescue basket is used, the full complement of RIT tools will remain mobile. This is extremely important in the event that the RIT must quickly deploy to another area such as another sector around the building, an exposure building, a subway tunnel, the other side of a wide-area building, or the floor above in a high-rise building.

As mentioned earlier, the type of additional tools added to the RIT staging area will depend on the complexity and severity of the fire scene. Gathering additional tools may not be a task for the original RIT, but rather for the additional RITs that are assigned to the rescue site. Even if the original RIT is activated after they have witnessed an obvious collapse, they will still have to recon the rescue scene with basic tools before moving in heavy hydraulic tools, shoring, etc.

The need for mobility of RIT tools is the reason that a rescue basket is preferred to spreading out a tarp and laying out the tools on it. Although one advantage of using a tarp is that it can be color coded or stenciled to identify the RIT, this same process can be applied to the Stokes basket by stenciling it and the search rope bags.

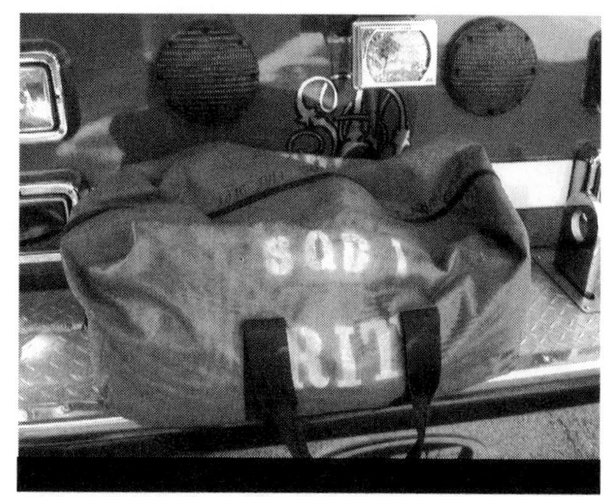

Fig. 3–14 Equipment Labeled as RIT Equipment Will Provide Scene Identification of RIT Staging Area

4

Firefighter Survival Rules

Firefighter survival can be broken down into self-survival, partner survival, and team survival. Whether helping yourself, your partner, or working as a team to get out of harm's way, survival seems to come down to these five basic rules:

RULE #1: Know when and how to call for help: "Mayday!"

RULE #2: Never give up!

RULE #3: Think outside of the "box."

RULE #4: It is advised not to share your SCBA air.

RULE #5: Keep control of the fire if at all possible.

A discussion of these five rules follows.

RULE #1: Know When and How to Call for Help: "Mayday!"

The minute a firefighter senses that they may be in personal danger, they must call for help immediately. This lesson was shared by Worcester District Fire Chief McNamee after the December 3, 1999, Worcester Cold Storage Warehouse fire.

Fig. 4–1 District Chief Michael McNamee, Worcester (MA) Fire Department (Courtesy of Associated Press, David Kamerman, December 1999)

As firefighters, we are trained to assist and help others who are in desperate trouble. Whether dealing with fire, collapse, flooding, ice rescue, sickness or trauma, the mentality is always, "How can we help others?" It is this same training and mentality that can get firefighters and many other rescuers into grave danger by not calling for help for themselves soon enough. The need to call for help can range from being in a minor entanglement that simply requires a partner to reach over and snip a wire to an SCBA malfunction that leaves the firefighter with no air at all and requires a full RIT response.

To ensure that firefighters call for help immediately, a fire department might consider using varying types of communication such as the following.

Urgent or emergency radio transmission. Situations that require this type of communication call for radio silence and for firefighters in the immediate area of the problem to handle it. Note the following situations that fall into this category.

1. Minor SCBA air leak
2. Minor entanglement
3. Firefighters exiting building on SCBA low-air alarms
4. Investigation of an activated PASS alarm
5. Investigation of a firefighter who is unaccounted for

Mayday radio transmission. Situations that require this type of communication call for radio silence and for firefighters in the immediate area of the problem attempt to assist. They additionally require the activation of the RIT.

1. Lost firefighter on SCBA low-air alarm
2. Difficult entanglement
3. Complete loss of SCBA air

4. Lost firefighter due to a confusing room configuration, vast room size, or collapse cutting off exit
5. Physically trapped firefighter

Emergency verbal communication measures to be used in these situations when calling for help are:

- Calling out "Mayday! Mayday! ... Engine #4" ... if radio equipped.
- Activating the emergency button (if provided), if radio equipped.
- Calling out for help from partner or other companies that might be nearby.
- Utilizing a hand-held flashlight as a signal in order to be seen by rescuers.
- Striking the floor, walls, columns, or any object that will transmit noise loudly throughout the building.
- Breaking a window(s) for noise and possibly to gain the attention of firefighters outside. (**Caution**: Breaking a window can worsen fire conditions and cause an even more desperate situation for rescue. First close any doors to seal the room and check for fire in the walls and ceiling if possible before breaking the window.) When heavy smoke conditions exist there will be times when it will be difficult for firefighters in the street to see a problem, so utilize a hand-held flashlight out the window to signal for help, or drop a piece of furniture or any contents out the window to attract attention.
- Activating the PASS alarm. Although wearing a PASS alarm should be mandatory for all firefighters who enter a fire building, the actual activation of the PASS alarm by a firefighter who is in distress, although advisable, should be discretionary. It is extremely difficult for rescuers and distressed firefighters to communicate with the victim when competing with the shrieking sound of an activated PASS alarm. If radio contact, face-to-face communication, or another method of communication (e.g., rhythmic or repeated striking of a floor, pipe, or wall) between rescuers and distressed firefighters is successful during the search and rescue process, activating the PASS should be delayed. If other forms of communication have proven ineffective, then the alarm should immediately be activated to facilitate rescue efforts.

The sanctity of a Mayday distress call

The sanctity of a Mayday distress call for a firefighter downholds the greatest priority over all other messages at an emergency incident. It must be realized that anyone at the incident may make a Mayday distress call for help. Once the Mayday distress call has been requested, it should not be repeated unless there is a second request because the first was not heard or there is a second (or another) incident where firefighters are down. If you then need to request priority clearance on the radio say "Emergency!" or "Urgent!" Confusion has resulted when the original Mayday has been called and then, upon reaching the victim, a firefighter or the RIT calls another Mayday. It is of utmost importance to announce when the victim is located, but do not use Mayday unless there is an additional problem such as the surprise discovery of multiple victims.

How to know when to "call for help"

Since firefighters are trained to survive in some of the worst conditions in terms of heat and smoke, their dependency on personal protective equipment can become too great. Crawling too far into a "no return" situation where the heat level has reached flashover conditions or going in too far and running out of SCBA air has resulted in both serious injury and death to firefighters. Dependency on personal protective equipment has thus given firefighters a false sense of security and sometimes delayed the "call" for help.

Many researchers have collected data revealing time/temperature curves, turnout gear thermal protection performance, and human pain tolerance in relation to feeling heat levels in a structure fire. Much of this research is based on the ears of firefighters being uncovered and exposed to heat and fire. The findings have been debated for years over the issue of whether the ears can really feel heat in time to warn a firefighter that it is time exit an area and whether or not the risk of any exposed skin is worth a burn. Whatever the debate is and whatever type of turnout gear and SCBA is being used, the fact is that we must know fire behavior, building construction, and the limits of our personal protective equipment. Our best defense against "getting in too far" and failing to call for help sooner during a fire is:

- **Fire behavior training.** Know flashover and rollover indications along with smoke behavior. Training and experience in understanding fire behavior never stops during a firefighter's career. Being able to "read" smoke to determine the amount of fire that is behind it can save a firefighter's life. Being able to distinguish a black "heavy push" bellowing with high heat behind it from a gray "lazy push" with an easy lift indicating little heat behind it will determine just how far the efforts to fight the fire and/or a search can progress. Knowing the height of the ceiling and the level and amount of heat at any given point is critical during interior operations. Any early radical change felt should require the firefighters to use water, ventilate, or retreat before progressing any further without overcommitting.

- **Building construction training.** It is also important to be able to match up the fire behavior training and experience with the knowledge of building construction. The danger of any increase in temperature in an unprotected wooden "balloon" frame house with small rooms will be a very serious indication of immediate danger. However, an increase in temperature in a non-combustible industrial building with skylights, fusible-link roof vents, and large open areas does not indicate the same thing. Not that changing fire conditions in such a non-combustible building should be taken lightly, but its reaction can be more predictable than that of the wooden "balloon" frame house when being attacked by fire.

- **Personal protective equipment education.** Firefighters need to be familiar with thermal protection performance levels, total-heat loss levels, and the construction of turnout gear. In addition to general familiarity with their turnout gear's thermal resistance and encapsulating performance, it must be realized that individual firefighters may have differing tolerances and reactions to varying heat-level exposures even though they are wearing the same turnout gear. Therefore, firefighters need to know how "they" will react in their particular turnout gear by having experienced fire scenarios in the turnout gear that they will actually wear to fires. They can acquire this knowledge by training in fire scenarios in full turnout gear. Training in "test gear" or "training gear" to avoid damage to frontline equipment is a questionable practice for recruit firefighters since the training gear may differ greatly from actual turnout gear in many ways, which would not allow the firefighter to realize what their personal limitations may be. In addition to this training, firefighters need to be familiar with normal and emergency operations of their SCBA, with special emphasis on PASS activation and deactivation.

- **Experience.** Unfortunately, experience cannot be obtained instantaneously, but some can be passed on from senior firefighters. The more experienced members of a department need to relate experiences that will prevent younger firefighters from "getting in too far." Younger, less experienced firefighters often learn independent facts about fire behavior, but lack the situational experience to put the entire puzzle together on the fire scene.

By way of example, suppose there is a residential house fire with an order having been given for a primary search. Before entering the door, firefighters can see there is heavy black "chunky" smoke with a heavy push of pressure behind it. An examination of the outside of the home reveals very few windows for outside ventilation and as firefighters try to move the line in past the threshold, heavy heat can be felt pushing down and out of the door. In this situation, experienced firefighters would recognize that although their turnout gear and SCBA may be offering protection at the moment, this as a dangerous situation requiring immediate ventilation above the fire and cooling of the fire by the hose stream. They would also realize that a retreat to the entry door would be urgent before flashover occurs, trapping the crew or the search and rescue team on the line.

- **Feel the conditions.** With fully encapsulated personal protective equipment, it is difficult to determine if the fire conditions are changing and, if so, just how quickly they are changing. From time to time, it is advisable to remove a glove and slowly raise a hand to feel the surrounding conditions. Admittedly, this is not an acceptable practice according to some authorities who have established safety, training, or equipment standards. However, it is better to take a slight risk in order to feel the progressive heat changes while moving further into a building or going above a fire to effect a search and rescue operation. The obvious result from not being aware of the progressive heat changes is to get caught in what retired FDNY Deputy Chief Vincent Dunn calls a "no return" situation when a flashover occurs. The following is a case study that demonstrates how quickly fire conditions can change and bring about a firefighter fatality.

On January 28, 1995, a firefighter lost his life in an effort to save a missing resident of an occupied manor home. The building was a turn-of-the-century three-story unprotected wooden "balloon" frame rooming house. The 19 residents were transient in nature and had occupied most of the building's 22 rooms. During the life span of the building, numerous additions and alterations had been made. At the time of the fire, there were four distinct sections to the building—the original house, an addition to the rear (where the fire originated), and two separated additions to the rear of the building. The building had pitched roofs, dormers, knee walls, spaces between the ceiling and the ridgepole, and many other concealed spaces that allowed for fire travel.

The fire originated in the second floor bathroom where it was ignited by a light fixture. As the fire grew, the balloon construction allowed it to travel horizontally above the second-floor ceiling. Once the fire reached the outer walls, it traveled vertically into the knee walls, ceilings, and attic space.

As the first engine, Engine #4, arrived, they reported smoke coming from the rear of the building and requested an additional alarm. Exiting residents informed firefighters that there was a resident unaccounted for and possibly trapped on the third floor. The initial response of two lieutenants and seven firefighters was to commit to search and rescue operations. Engine #4 entered the building from a rear fire escape at the third floor level to search for the missing resident. Engine #3's crew (the victim's assigned company) entered the building and climbed to the third floor using an interior staircase to also conduct search operations. Initially, smoke conditions were reported as not being a problem. Room doors were being forced opened as a preconnected $1^3/_4$-in. hoseline was being stretched up the stairs by one other firefighter. It is thought by investigators that by this time the fire had extended considerably into the attic knee walls, attic peak, walls, and ceilings. Then, (approximately four to five minutes after arrival) fire broke out of the concealed spaces, dropping heavy smoke and heat to the floor. This occurred just after the exterior ventilation of four third-floor front windows where fire was seen breaking out of concealed spaces. It was at this time that the third floor "lit up" in a flashover forcing firefighters to escape with the exception of the firefighter victim who was suddenly trapped in one of the boarding rooms. A witness reported seeing him at the window without an SCBA facepiece, calling for help. The victim was eventually found dead with full personal protective equipment, SCBA facepiece off, and PASS alarm not turned on.

Factors that can impede the "call for help"

Several factors can cause or lead up to a fatality of a RIT team member. The following is a discussion of some of the most common ones.

Poor fireground accountability. Serious delays in calling for help often are a result of not realizing that a firefighter is missing. Many NIOSH fatality investigation reports indicate one of the main factors leading to fire fatalities is the lack of a firefighter accountability system. It is the company officer's duty to know the location and tasks being performed by their company members, and it is the IC's duty to know the location of assigned companies. Incident commanders need to institute an incident management system that allows them to track the location and assignments of all companies on the incident scene.

Not wearing and/or activating "stand alone" PASS alarms. Injury and death investigations have revealed over time that, in many cases, victims had not turned on their PASS alarms. It is mandatory that all PASS alarms be turned on when entering a building requiring fire suppression operations. To assure that PASS activation occurs when firefighters enter a burning structure, the 1998 edition of NFPA Standard 1982 added the following requirements to address this problem[7]. (Note: the term *activation* is also used to refer to the actual use of the alarm in an emergency situation. However, in this case, it means to enable the alarm so that it can later be activated quickly if needed.)

Section 4-1.2 The mode selection device(s) shall be designed to provide automatic activation from the off mode to the sensing mode without the user setting the mode election device.

Section 4-1.2.1 Such automatic activation shall be permitted to, but not limited to, linked to activation of SCBA, linked to removal from storage or transportation positions, by pull-away tether to a fixed position, or remote activation.

It must be realized that not all departments are financially able to update equipment, even safety equipment, every time a standard changes. Therefore, many departments continue to operate with PASS devices that require the user to manually activate the device before entering a fire building. Only repetitive training and strict enforcement of operating procedures that mandate activation of PASS devices will ensure that the device is activated before the firefighter enters the building.

On November 6, 1998, two volunteer firefighters (from two different departments) died trying to exit a burning auto salvage storage building. Altogether, three departments arrived on the scene. For the sake of clarity we will refer to them as Department #1, Department #2, and Department #3. Arriving on the scene, they found a metal-pole building with light smoke showing. The Chief of Department #1 assumed the position of IC and discussed the possible origin of fire with the owner of the structure. The IC then decided to ventilate by ordering a firefighter to open one of two small roll-up garage doors on the north side of the structure. He proceeded to the southwest corner of the structure where he ordered the owner to tear off metal exterior wall panels with a fork-lift. Once ventilation was completed, three members of Department #2 (Chief, Assistant Chief, and firefighter) and three members of Department #3 (Captain, Lieutenant, and firefighter) advanced two $1^3/_4$-in. lines through the front door of the building, which was filled with light smoke. As firefighters proceeded to the rear of the structure to determine the fire's origin, heavy black smoke collected below the ceiling, and small flames trickled over the ceiling's skylights. Approximately 80 ft. inside the structure, firefighters found what they believed to be the seat of the fire and began to apply water. As fire-fighting activities proceeded, firefighters transferred the lines to other firefighters because the low-air alarms on their SCBA were sounding. Approximately 11 minutes into the attack, the IC ordered both crews to exit to discuss further strategy. As the crews began to exit, an intense blast of heat and thick, black smoke covered the area, forcing fire-fighters to the floor. The Chief (victim #1) and Assistant Chief from Department #2 were knocked off their hoseline, and their SCBA low-air alarms began to sound as they radioed for help and began to search for an exit. The two departed in different directions, and the Assistant Chief eventually ran out of air and collapsed. He was found immediately and assisted from the burning building. As firefighters pulled the unconscious Assistant Chief to safety, the Lieutenant (victim #2) from Department #3 re-entered the structure to search for the Chief. During his search, the Lieutenant ran out of air, became disoriented, and failed to exit. The Lieutenant was discovered equipped with a PASS; however, the system was not activated. The Chief was known to have entered the structure without a PASS device. Additional rescue attempts were made but proved to be unsuccessful. The body of the Chief, was eventually located.[8]

Misguided training. Delaying a call for help will result in a loss of precious time and the consumption of SCBA air. Firefighters who are lost will generally rely on their training to attempt to retrace their search path, follow a wall, locate a window, or even "buddy breath" SCBA air, thinking they will find their way out without help. Even when these methods are not successful, firefighters generally do not call for help because they believe they will have a safe outcome just as they have experienced in training exercises. Sadly, those training exercises have usually never included how to make an emergency call for help. This method of training firefighters to rely only on themselves or a buddy must change now.

When probationary firefighter recruits are being trained, training officers must educate and instruct each firefighter how to recognize when they are in trouble, when to call for help, and how to call for help. Firefighter training that concentrates on disentanglement, rescue, shared SCBA air methods, and other emergency techniques, does not consistently include a definitive and mandatory call for help, which paralyzes firefighters in actual distress situations because they have no training on how to call for help. Advanced SCBA training and firefighter rescue programs must also emphasize to veteran firefighters when to call for help. (There is an additional immediate need for training that will improve skills in escaping confined spaces by reducing the SCBA profile, removing the SCBA, and using disentanglement techniques.)

Lack of experience. Firefighters who lack experience may not even be able to identify when they are in trouble and that they should be calling for help. Take for example two firefighters who almost died shutting down a sprinkler valve in the basement of an old mill-constructed factory. After the fire was extinguished and overhaul was to start, they followed their orders by entering the basement, which was, like most factory basements, congested with boxes, pallets, and loose wire and now contained about 6 in. of water throughout. However, their SCBA cylinders were only half full. As they proceeded, the heavy, wet smoke conditions allowed them to see only with hand-held flashlights. Suddenly, the first of two SCBA low-air alarms activated. At this point they were committed about 50 ft. into the basement. Soon afterward, the second SCBA alarm activated, and the first SCBA ran out of air. Only then was a desperate radio call for help made. It was fortunate that rescuers got to the firefighters in time. However, there have been similar incidents where this was not the case.

Fig. 4-2 NIOSH Case Study Report 98F-32—The Burning Auto Salvage Storage Building Had Presented Difficulties with Not Only Deteriorating Fire Conditions, but also a Complicated Interior Layout and a Large Search Area (Courtesy of Goldboro News-Argus)

Denial of the situation. The same type of denial discussed in chapter 1 in relation to the rescue of another firefighter can be witnessed in the context of self-rescue. While wrestling with the problem, consuming air, contending with fear and confusion, there is a sense of denial that "this is not happening to me!" while the clock ticks away and precious time is wasted.

Self-pride. The dynamics of the fire service are such that we take so much pride in what we do as a fire department, a fire company, and individually that we create a fear of retribution from fellow firefighters in the firehouse if we make almost any type of mistake. We do whatever it takes to preserve and protect our pride and to not confront what seems to be the never-ending ego bashing from our peers. Hence, the thought of calling for help is perceived as ammunition for our peers to use to criticize us. Unfortunately, in this case, being too proud, too paranoid, or being afraid of criticism can cost us our lives. This is where training is needed to reinforce good instincts, habits, and procedures and to reduce the fear of peer criticism.

Poor radio usage. As previously discussed, radio communications often fail on the fireground and even more so during an urgent or Mayday situation. An unheard or misunderstood Mayday distress call transmission will cause a serious delay in rescue or a rescue attempt in the wrong part of the building. It is important that firefighters use the radio correctly during routine operations. It has been commonplace to experience transmissions that are not complete because the transmit button is not pushed in time or is released too early. Attempting to transmit through the SCBA is another feat requiring training on how to use the speaking diaphragm or just to speak slowly and calmly. As in any radio message during the heat of the moment, it should be kept brief and to the point. Some departments choose to use numerical codes while others prefer plain English to communicate on the radio. Both procedures have their merits, but either case can be ineffective if the firefighter chooses to bury the initial transmission with extensive follow-up details, broken conversation, or numerous repeated messages. One example of how a company officer could account for other firefighters might be:

"Squad #1 to Squad #1A, OK?" or

"Progress Report?" or

"PAR (Personnel Accountability Report)?"

In reply say,

"Squad #1A OK, Sector A, second floor search."

That simple "OK" is with reference to the accountability and condition of the firefighters. Using numerical codes, the way a company officer might account for other members could be:

"Squad #1 to Squad #1A, 10-10?" or

"Squad #1A, Sector A, second floor search."

Both methods not only account for the presence of all firefighters but also their location and job task in a concise and clear transmission.

RULE #2: Never Give Up!

In the context of firefighter survival, "Never give up!" means to keep trying every different means of escape you have experienced through training. It is important for every firefighter to know that they should never lie down and stop trying to escape if they can physically keep moving. Firefighter survival training is training that must provide a menu of options for survival such as breaching through walls to escape, disentangling from wires, or rappelling from an upper floor when escape is cut off by collapse or fire. If conditions allow, the trained firefighter might be able to try a number of different survival options to attempt to escape the situation or at least survive until help arrives.

The mind is a very powerful part of firefighter survival. It takes a strong drive for the firefighter to keep presence of mind, use the learned survival skills, and "Never give up!" In some cases, firefighters have virtually given up in their survival attempts when they were entangled in wires or caught in a smoky room with lightweight constructed walls. Both of these situations were escapable had the firefighters been adequately trained and able to keep their presence of mind. As long as the firefighter can refer to trained survival skills, there will be a stronger drive to survive without entering a state of panic or simply giving up.

On April 11, 1994, at approximately 2:20 a.m., two firefighters lost their lives on the 9th floor of a high-rise apartment building. To reinforce the need for survival skills and tools, we will focus on only one of the two firefighter victims. The officer (victim #1) had placed one of his firefighters (victim #2) in a safe area (either an apartment or closed stairwell) due to victim #2 having SCBA problems. As the officer advanced back down the hallway toward the fire in an effort to locate his other firefighter, fire conditions rapidly changed which resulted in his death. It was unknown to victim #2 that his officer was in trouble, even though he had made four unsuccessful radio transmissions to contact his officer by radio. Victim #2 then re-entered the hallway and became entangled in cable television wire that had fallen from a secured position on the wall near the ceiling. (The heat from the fire had melted the plastic encasements on the cable causing it to fall across and hang downward in the hallway.) When victim #2 became entangled, he was 9 ft. from the protection of the stairwell door. When found, victim #2 was face-down with cable wires wrapped around his SCBA, his upper torso, and legs. Although the SCBA air cylinder valve was only partially opened, the regulator was properly opened, and the cylinder was empty of air. Even though the SCBA did not pass certification standards, it was deemed not to have contributed to the firefighter's death.[9] What then might have led to his death? It was conjectured that as the firefighter was stretched out facedown, entangled with his facepiece still in place, he possibly concluded that he could not escape his entanglement.

It is not our purpose to assume that victim #2 gave up in any way, but rather to learn from this situation. In entanglement scenarios, one must be able to react almost instinctively to disentangle as a result of prior training and to remember to stay calm and call for help. These two factors will increase your chances of survival and reinforce the fact that you should "Never give up!" Additionally, prior training in any self-survival scenario will increase the likelihood of your survival in an actual incident.

RULE #3: Think Outside of the Box.

To think outside of the "box" is, in one sense, a way to describe how to think beyond normal limits, but in firefighter survival it is a thought process of ingenuity and innovation for survival. Thinking outside of the "box" when trying to escape a room or building may require some unconventional moves, but they might save your life. In 1973, the Forum Cafeteria fire in Chicago resulted in a bowstring-truss roof collapsing and killing three firefighters. In addition to the fatalities, there were a number of firefighters trapped on a second floor mezzanine under the fallen roof. Some of the trapped firefighters were able to slowly crawl around because the cafeteria tables and chairs provided a void space by holding up much of the fallen roof. Although not able to reach the staircase, one of the firefighters finally reached a wall and began to search for a window or door to open. Upon doing so, he found a dumbwaiter shaft, which was used for transporting food and dishes between the first floor kitchen and the mezzanine. The shaft was small and confining, but fresh air and light was found coming up from the bottom. As the burning roof above them collapsed, five firefighters removed their turnout coats and slid down feet-first to

Fig. 4–3 Forum Cafeteria Fire in Chicago 1973 Moments Before the Bowstring Truss Roof Collapsed (Chicago Fire Department)

the bottom, one at a time. This is one of many examples of having to think of alternate methods of escape.

RULE #4: It Is Advised Not to Share Your SCBA Air.

As mentioned earlier in chapter 1, "Never say never; never say always," but as a rule, it is advised not to share your SCBA air with a victim unless they are entangled, pinned, or trapped so that sharing of SCBA air is the only alternative. If the victim is free to move about and can walk, crawl, or be dragged, it is advisable to not share your SCBA air. The idea of sharing SCBA air is something that we must look at carefully. Just because there is an SCBA malfunction or loss of remaining air does not mean we have to "buddy breathe" or share air. Imagine a smoky fire on the second floor of a house where you and your partner have just reached the second floor hallway to advance toward the back bedroom. While crawling about 10 ft down the hall, your partner's SCBA fails (for whatever reason), and he indicates that he is trouble. Would it be better to remove your helmet and hood so you could pull off your SCBA facepiece to share it with your partner or aggressively grab and pull your partner down the staircase to safety? There is no question that pulling your partner to safety down the staircase is the fastest and safest way to go for both of you. In most cases, there are reachable doors, windows, and staircases that will lead to safety to which the victim could be pulled. Granted, the victim might breathe in some smoke and receive a few bruises while being pulled, but the rescuer will assure that the rescue is a success by not giving away any precious SCBA air. In the case of a missing firefighter who is already in a state of fright and is either low or out of air, you can count on not getting your SCBA facepiece back at all if you attempt of to share it. The bottom line is that the situation has the potential of going from bad to worse if you share you SCBA facepiece for the following reasons:

- The most common and quickest way to share your SCBA air is to give your partner your SCBA facepiece. Although this may be true, not only will you loose whatever precious remaining air you have while passing it back and forth, but there is also no guarantee that you will get the facepiece back. Personal experiences have shown that when firefighters are endangered and down to their "last" breath, self-preservation will take over and they will not give your facepiece back and may even reach to pull it off your face. Now you have gone from a position of rescuer to being a victim yourself.

- There are other options in sharing SCBA air such as equalizing the air from the rescuer's SCBA to the victims. Again, this will take time. The time that it will take to connect the equalization line to transfer the air might be the time it would have taken to get the victim to safety. When equalizing SCBA cylinders, the rescuer will transfill 50% of the remaining air in their SCBA cylinder to the victim. The rescuer who had only 10 minutes of SCBA air time left after searching will now have only 5 minutes remaining. That might not be enough to get out.
- There are other SCBA systems that are designed for "buddy breathing" shared-air operations approved by the NFPA. Again, the rescuer must determine if it is worth the time and risk to share air or if it is more expedient to just get the victim out of the building. Training in how to use the integrated "buddy breathing" systems and when to use the systems is most important.

There may be a time when both partners are mobile and will have to share SCBA air as they are calling for rescuers and searching for a way out of a large, smoky building. Similarly, there may be a time when you must share SCBA air with a firefighter(s) who is entangled, pinned, or trapped in order to save the victim's life long enough to evacuate them to safety. These last two scenarios are examples of why it is said that there is "an exception to every rule." Hence, training, experience, and common sense are the best "rules" for judging whether the situation warrants the practice of sharing SCBA air or not.

RULE #5: Keep Control of the Fire if at All Possible.

Whether in the case of firefighter self-survival or rapid intervention rescue, the fire must be kept under control or at least "in check" if at all possible until the firefighter(s) are rescued. When this rule is violated, as when all hands are working on the rescue, the fire will take ownership of the building quickly, resulting in the following problems:

1. Increasing the risk to the rescuers and victim.
2. Worsening the fire conditions with increased smoke, heat, and threat of fire spread; thereby reducing the chance of a successful rescue operation.
3. Reducing the span of time to effect a successful search and rescue operation.

4. Assigning companies away from the rescue to continue the efforts to fight the fire (As difficult as this might be to do, it is a strategic move that may be highly responsible for a successful firefighter rescue.) This is one of the primary advantages of having a department rule or operating procedure which dictates that the RIT be deployed to be able to focus on the rescue while other companies remain in the efforts to fight the fire to protect the search and rescue efforts.

Violation of this rule can happen in three ways:

1. Larger fire departments with adequate staffing allowing companies to abandon their firefighting positions to respond to the search and rescue.
2. Smaller town fire departments that do not call for additional staffing via mutual aid to provide a RIT. Without a RIT during a Mayday situation, the only companies available to respond are the companies fighting the fire.
3. A fire department that never had control of the fire in first place and now must respond to a firefighter Mayday distress call. In such cases, the rescue is usually unsuccessful and the fire grows in magnitude.

Whether the control of the fire is lost due to a fast-spreading fire, poor construction, inadequate staffing, poor tactics, severe weather, equipment breakdowns, Murphy's Law, or personnel breakdowns (firefighters missing, lost, or trapped), disaster is sure to follow. An article from the Columbus Monthly magazine, "The Murder of John Nance," relates how Columbus (OH) Firefighter John Nance became trapped in the basement of a four-story, 11,500 square-foot, downtown commercial building:

> *"I think I stepped on him (the victim, Firefighter John Nance) on my way back out." Brining (Firefighter assigned to Rescue #2) continues. "But my alarm bell went off. When you get low on air you've got about two minutes left when the bell starts to ring. And I knew I was deep into the building. Sometimes two minutes is not enough. By the time I got to the back of the building, I was sucking the mask to my face. I was out of air." Nance was out of air too. The building was getting hotter and smokier by the second. Fire began to show in the upper floors. "Conditions were beginning to look as though they might soon be ripe for a flashover, or perhaps even a backdraft explosion," recounted Linsey (Battalion Chief).*[10]

In spite of every courageous effort, Columbus firefighters had to abandon the rescue attempt and evacuate the building because the fire had taken control of the building.

5

Firefighter Survival Emergencies

Entanglement Emergencies

During fireground operations, an entanglement emergency occurs when a firefighter has become tangled in debris that prevents the freedom of movement, thus creating a potentially life threatening incident. Debris that can create an entanglement emergency are:

- Television and computer cable
- Electrical and telephone wiring
- Suspended ceiling grid and wiring
- HVAC flexible tube and aluminum ductwork
- Draperies and blinds
- Furniture (e.g., burned mattresses with exposed bed springs)
- Commercial store fixtures (e.g., clothing racks and shelving)
- Industrial machinery (e.g., wire, handles, levers, etc.)
- Building components (e.g., banisters, railings, collapsed debris, etc.)
- Other materials and objects

Fig. 5–1 Simulated Firefighter Entangled in Television Wire (Kolomay)

SCBA Basic Training for Disentanglement Survival Training

Basic SCBA donning and removing techniques

In an effort to learn many of the possible survival techniques to escape entanglement emergencies, it is most important to be expertly proficient in the design, functions, capabilities, and limitations of the SCBA. It is important to repeatedly drill on the basic methods of donning and removing the SCBA. Drills should be conducted to include a variety of circumstances and locations in which firefighters will encounter a need to don their SCBA. At a minimum, this should include donning the SCBA while being seated in the apparatus, picking up and donning the unit from the ground, and removing the SCBA from an apparatus compartment and donning the unit. Other skills that should be included in every drill include making sure all of the harnesses are properly donned and tightened and that the cylinder air pressure is at maximum. Additionally, the low-air alarm and PASS pre-alert (if integrated) should be pretested to ensure that they are fully functional.

The proficiency of any firefighter in using the SCBA is based on the intensity of the training experienced. Examples of intensive SCBA training that could improve a firefighters' proficiency would be as follows:

- Donning the SCBA while being timed
- Donning the SCBA using several different methods (e.g., "coat method" or "over-the-head method")
- Donning the SCBA in darkness
- Donning the SCBA while being timed and in darkness while listening to orders via radio (and repeating those orders after the SCBA is completely donned)
- Activating the emergency button on the PASS alarm in darkness with gloves on
- Deactivating the PASS alarm on an unconscious firefighter in darkness with gloves on
- Removing and inserting a mask mounted regulator in darkness with gloves

Repetition of such intense SCBA drills will develop a firefighter's skills into unconscious habits allowing any firefighter to feel confident in SCBA usage and better equipped to handle SCBA and/or entanglement emergencies. Many SCBA manufacturers offer varying features such as quick fill, "buddy breathing" capabilities, and varying regulator and facepiece designs. In an effort to ensure common safety features and gain some consistency in SCBA design and use, NFPA Standard 1910 was developed. This standard establishes and standardizes many SCBA design features.

Minor entanglement emergencies

A minor entanglement is defined as an entangled firefighter that is caught or tangled briefly, only requiring a shift in body or SCBA position, a change in direction, or the easy removal of the entanglement by hand. When involved in any entanglement situation in which the firefighter is tangled and prohibited from moving forward, DO NOT keep pushing forward to break the entanglement. This could possibly cause knotting of the wire, tightening the entanglement, and even collapse of acoustical ceiling, shelving, or other similar obstacles. There are several survival techniques that can be used to escape such entanglements.

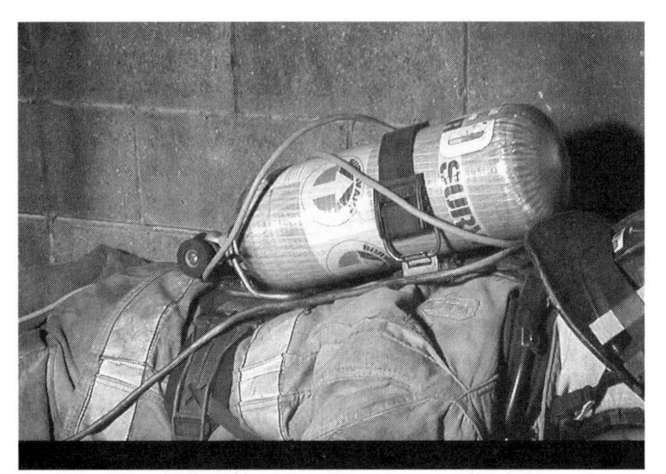

Fig. 5–2 Entangled Firefighter Crawling Forward is Caught by Cable, Entangled, and Can No longer Move Forward (Kolomay)

Fig. 5–3 Entangled Firefighter Moves Backward to Release Tension on Entanglement (Kolomay)

Fig. 5–4 Entangled Firefighter Rolls Attempting to Disentangle and Remove Entangling Cable (Kolomay)

Back up and turn method. The entangled firefighter should:

1. Notify partner of a minor or serious entanglement so they do not become separated and so that they can work as a team to quickly disentangle.
2. Attempt to back up enough to release whatever entangled tension there is, and slowly turn to the left or right as needed. Once turned, attempt to back up again with the hope of releasing the entanglement. If the cable is not fully released, attempt to roll to the right or left side to disentangle and remove the cable.

Fig. 5–5 Loosen the Right or Left Shoulder Harness and Waist Harness (Optional) (Kolomay)

Reduced SCBA profile method. In the event a firefighter cannot move forward due to being in a confined, irregular space or entanglement material, certain situations might require the shifting of the SCBA to reduce the profile of the firefighter to allow passage. Shifting the SCBA can be achieved while lying, kneeling, or standing depending upon the situation. The entrapped firefighter should:

1. Loosen the right or left shoulder harness and waist harness (optional).
2. Shift the back of the SCBA frame and cylinder to the right or left by pulling the waist harness, pushing on the appropriate shoulder harness, and using some twisting momentum to slide the SCBA off to the side. The SCBA can be shifted to the left or right depending upon the confined space or entanglement situation. Some procedures outline the need only to shift the SCBA to the right side so that the left hand is free to do such things as hold on to a regular breathing tube. However, experience with various types of confined spaces and entanglement situations have demonstrated that the firefighter needs to know how to shift the SCBA in either direction in order to reduce a firefighters profile.

Fig. 5–6 Shift the SCBA Back Frame and Cylinder to the Right or Left by Pulling the Waist Harness, Pushing on the Appropriate Shoulder Harness, and Using Some Twisting Momentum to Slide the SCBA Off to the Side (Kolomay)

3. Move through the confined area with the SCBA in the reduced profile position.

Fig. 5–7 Firefighter Moving Through a Confined Space with SCBA in a Reduced Profile Position (Kolomay)

Serious entanglement emergencies

A serious entanglement is defined as an entangled firefighter who is caught or tangled requiring one or any combination of the following actions to escape.

Assistance from a partner(s). A firefighter partner is needed to assist with disentanglement, tools, communication, shared SCBA air, and moral support. The firefighter partner hopefully will be in a better position to recognize and release the entangled victim. Most importantly, the victim's partner must avoid becoming entangled in the same material when assisting the victim. The procedure for this is as follows:

1. A Mayday distress call must be given immediately once it has been quickly determined that the victim is involved in a serious entanglement situation.
2. The victim must conserve SCBA air and not struggle. The firefighter partner must calm the victim in an effort to slow down the breathing rate.
3. The firefighter partner must work on the disentanglement as the victim remains still.
4. The firefighter partner must be prepared to direct rapid intervention personnel to difficult release points, needed tools, and assist with shared-air operations if needed.

Disentanglement with wire cutters and/or knife. One of the key aspects of a firefighter carrying personal tools is to avoid carrying a "tool box" of so many tools and gadgets that the tools themselves become a problem rather than a solution. Not only is there concern for additional weight in carrying too many tools but also a concern about entanglement problems presented by protruding tools. One of the most important personal survival tools has been found to be 6-in. or 8-in. wire cutters or a versatile knife

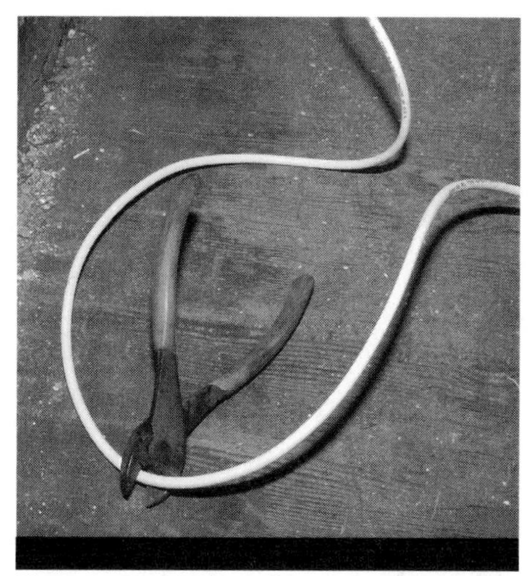
Fig. 5–8 Coaxial Cable and 6-in. Side Cutters (Hervas)

capable of cutting through wire. In either case, the quality of the tool will determine its worth. Cheaper tools have a tendency to fall apart, bend, or dull, all of which are critical detriments when your life depends on them.

Another important concern is the placement of this tool on the firefighter. The tool should be easy to reach when in full turnout gear, with SCBA on, lying down, and entangled. In some cases, depending upon design, turnout gear pockets will work as an adequate place for tool storage. If possible, placement is optimum somewhere near the collar. This can be accomplished by fastening the tool on the turnout coat or to a radio harness near the remote microphone. With or without the assistance of a partner, the chances of surviving an entanglement situation are increased if the victims can extricate themselves by cutting away and releasing the material, thus performing a self-rescue.

Rescue by the RIT. Serious entanglement may require the RIT to enter the rescue area with a thermal imaging camera, cutting tools, and additional SCBA air, to assist or perform the rescue. Once locating the entangled firefighter, the RIT officer must do the following.

1. Check the condition of the victim and the SCBA air supply.
2. Keep the victim calm and breathing slowly while supplying additional SCBA, if need be.
3. Determine the seriousness of the entanglement, assess how difficult the rescue will be once the victim is released, and be aware of the fire conditions at all times.

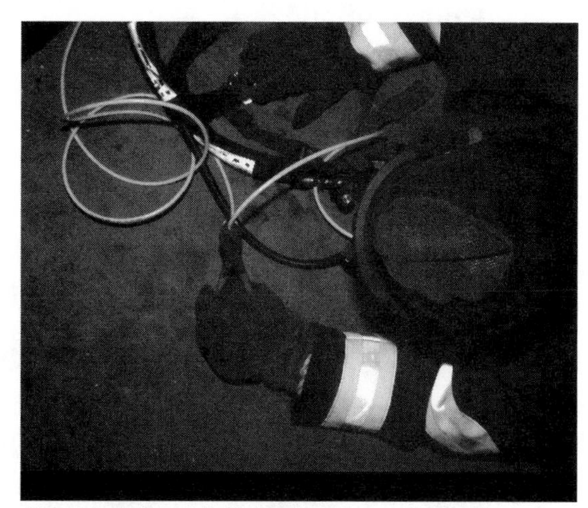
Fig. 5–9 A Potentially Serious Entanglement Situation (Hervas)

SCBA Self-Removal Techniques for Disentanglement

For an entangled firefighter, the removal of the SCBA to disentangle for escape is a last ditch effort for survival. Indications for SCBA removal are:

- The entanglement material cannot be released from the SCBA to allow the victim to escape to safety
- The entanglement material and/or a confined space will not allow the victim to escape due to the size or profile of the SCBA

General procedure for emergency SCBA removal while crawling

1. Attempt to roll onto the SCBA cylinder on whatever side will allow the easiest removal of the unit, while considering the possible restrictions of walls, collapsed material, and entanglement material.
2. Extend and remove the "high side" shoulder harness.
3. Release and remove the waist harness.
4. Roll out of the SCBA lying chest down; stay "on air" keeping the facepiece on.
5. Move the SCBA forward and collect the harnesses on the back of the SCBA frame.

Fig. 5–10 Extend and Remove the "High Side" Shoulder Harness, Then Release and Remove the Waist Harness While Still Breathing SCBA "On-Air"

6. Depending on the fire conditions, collapse conditions, distance to safety, and remaining SCBA air, the determination to fully re-don the SCBA should be determined by the firefighter or victim. To reduce the chance of additional entanglement of a loose SCBA during an escape, place the shoulder harness (minus the attached breathing line) over the head and cradle the SCBA under the right arm.
7. MOVE!

Fig. 5–11 Collect the Harnesses onto the Back of the SCBA Frame and Move Forward to Escape (Kolomay)

General procedure for emergency SCBA removal while kneeling or standing

1. Release the waist harness buckle.
2. Grasp the appropriate shoulder harness with one hand and loosen the other shoulder harness, and then remove your arm.
3. Turn around to face the SCBA and disentangle (while staying "on air" keeping the facepiece on).
4. Depending on the fire conditions, collapse conditions, distance to safety, and remaining SCBA air, the determination to fully re-don the SCBA is questionable. To reduce the chance of additional entanglement of a loose SCBA during an escape, place the shoulder harness (minus the attached breathing line) over the head and cradle the SCBA under the right arm.
5. MOVE!

Fig. 5–12 Emergency Escape (Place the shoulder harness, minus the attached breathing line, over the head and cradle it under the left arm so the firefighter will have access the high pressure cylinder valve, universal air transfer coupling, and belt-mounted regulator and air valves, if applicable.) (Kolomay)

SCBA Shared-Air Emergency Methods

Shared-air emergency methods (more commonly known as "buddy breathing") are acts of transferring the SCBA air from a rescuing firefighter to a victim firefighter who is low or out of air. Buddy breathing without approved equipment is not an action recognized by the NFPA, NIOSH, or SCBA manufacturer's literature. However, buddy breathing methods have been included in firefighter training for the past 25 years in firefighting SCBA courses throughout the nation. With the steady improvement and modernization of SCBA, intrinsic components have been added allowing the equalization of SCBA air cylinders with a "quick fill" feature. The sharing of air with a breathing line between SCBA has been recognized as being "approved" methods by national standards.

"Unapproved" shared-air methods, e.g., use of a common facepiece, that constitute last ditch efforts for survival, might be needed during the following situations.

- The SCBA being used is not modern enough to have integrated shared-air features.
- The SCBA worn by the victim and rescuer are not of the same manufacturer or are of different generations (e.g., one SCBA has a "quick fill" feature, and the other does not).
- Due to smoke and/or heat conditions, it is not possible for the firefighters to go through the process of deploying a shared-air line or equalization line.
- The firefighter victim's SCBA has been completely removed (e.g., in an attempt to perform disentanglement). The use of any integrated shared-air feature would be impossible since the SCBA air must be shared using only the air from the rescuing firefighter.

It is important to restate survival Rule #4 from chapter 4, "It's advised not to share your SCBA air." If the victim is free to move about and can walk, crawl, or be dragged, it is advisable not to share your SCBA air. During firefighter rescue operations (with a planned rescue) involving a firefighter partner or as the result of your actions on a RIT, it is recommended that you not share SCBA air with the victim. Such action can generally slow down the removal process, cost the RIT member valuable air for the rescue operation, and the victim might not be so willing to "share" the rescuer's air.

Within the context of firefighter survival as opposed to rescue, if among a team of firefighters there is an SCBA "low air or no air" emergency, shared-air methods are appropriate when the victim is:

1. Seriously entangled
2. Pinned (e.g., trapped under a beam)
3. Trapped in a room or void area due to a collapse
4. Lost and in search of escape
5. Seriously injured and incapacitated

Although the use of shared-air methods has already been recognized as an inadvisable action in chapter 4, it may be considered as a last ditch effort to survive.

Common facepiece method

The common facepiece method involves the removal of the SCBA facepiece with "good air" from the firefighter and passing it quickly to the firefighter in need of air.

The recommended procedure is:

Fig. 5–13 Common Facepiece Method (Kolomay)

1. Stretch the facepiece head harness and straps back over the lens to allow for a quick, easy, and sure seal when passed back and forth.
2. The firefighter with "good air" who is passing the facepiece, MUST keep a secure hand on the facepiece at all times in order to keep control and not lose it.
3. "Share" the air. Try to establish a "three-breaths-each" pace, then pass it back to the firefighter who is sharing the air.
4. When passing the facepiece back and forth, guard against the loss of air.

Caution: A strong Caution must now be restated concerning this method. There is a risk that the firefighter victim who is out of air may not give the facepiece back, due to a strong instinct for survival that can take over.

Common MMR (mask mounted regulator) method

The Common MMR Method involves the passing of the regulator with "good air" to the facepiece without air and back again. This procedure may not work between different manufacturer's SCBAs or even between different generations of SCBA manufactured by the same company. The twisting and positive locking of the MMR into the facepiece will vary in difficulty from one manufacturer to another, making this maneuver very difficult in some cases. The advantage with this method is that the facepieces remain on the firefighters, allowing for more facial protection from heat, debris, and some filtering of smoke. In addition, as the MMR is passed, the flow of air can be stopped to minimize its loss between firefighters.

Caution: It is important for the firefighter who is sharing the air to *never* let go of the MMR. The recommended procedure is as follows.

1. Reach for the MMR, release the lock, and twist.
2. Attempt to stop the airflow before removing the MMR from the facepiece.
3. Insert the MMR into the victim's facepiece.
4. Share three breaths with the victim.
5. Remove the MMR from the victim's facepiece.
6. Return the MMR for three breaths.

Common breathing tube method (belt mounted regulator SCBA only)

The common breathing tube method allows the firefighter who is out of SCBA air to insert the facepiece breathing tube into the side of the facepiece of the firefighter with good air. The positive pressure from the "good air" facepiece will supply both firefighters with adequate air. The recommended procedure is as follows:

1. The firefighter in need of air disconnects the breathing tube from the belt-mounted regulator.
2. The breathing tube is then inserted into the side of the good air facepiece. It is IMPORTANT that the breathing tube be inserted into the nose cup. If outside of the nose cup, it will become difficult to obtain air.

3. The firefighter with "good air" will have to attempt to pinch the facepiece around the inserted breathing tube to keep the best seal possible to minimize any loss of air.

Fig. 5–14 Common Breathing Tube Method (The firefighter victim on the rescuer's left has positioned his breathing tube inside the rescuer's facepiece. The rescuer's positive pressure facepiece airflow will supply air to the victim continuously.) (Kolomay)

Common regulator method (belt-mounted regulator SCBA only)

The common regulator method is designed for belt-mounted SCBA regulators that are typically earlier generation units compared to the mask-mounted regulator (MMR) units. This method allows for the victim to position the end of the SCBA facepiece breathing tube at the air portal of the partner's good air regulator and alternate the placement of the two breathing tubes after every three breaths. The recommended procedure is:

1. The firefighter in need of air disconnects the breathing tube from the belt mounted regulator.
2. The breathing tube is then inserted into the side of the "good air" facepiece. It is important that the breathing tube be inserted into the nose cup. If outside of the nose cup, it will become difficult to obtain air.
3. The firefighter with "good air" will have to attempt to pinch the facepiece around the inserted breathing tube to keep the best seal possible to minimize any loss of air.

Fig. 5–15 Common Regulator Method (Kolomay)

RIT Shared-Air Methods for Rescue

One of the difficulties found during firefighter rescue operations involving extensive disentanglement using pneumatic, hydraulic, and cutting tools, breaching walls and floors, and using rope for rescue is that each of these maneuvers can take a long period of time. Time can become as great an enemy as a weakened building or the fire itself. The limited amount of SCBA air is a problem for rescuers concentrating on a rescue operation. In repeated training scenarios, it was found that when low-air warning alarms sounded, the rescue operation started to fail. During two years of extensive RIT training with the Chicago Fire Department, more than 380 firefighter rescue scenarios were conducted with 60 truck companies in various vacant buildings. RIT work time averaged between 15 and 17 minutes with 30-minute SCBA air cylinders and, if the rescue had not been completed by that time, the following issues started to arise.

- The RIT firefighters did not want to leave the rescue area.
- If RIT firefighters did leave, it became difficult to understand just what task had been left unfinished (e.g., cribbing, supplying rope, or moving a victim).
- Communications became difficult between RIT firefighters trying to talk over a PASS alarm, SCBA low-air alarms, radio messages, and other surrounding noise.
- The anxiety and fear of leaving the victim and failing the rescue was greatly increased.
- Replacement RIT firefighters did not always know what rescue technique was in progress and started a different operation resulting in mass confusion and eventual failure.

It has been important for any units backing up the RIT rescue operation to provide the RIT with additional air. If additional SCBA air is to be supplied, it is important that whole SCBA be exchanged with RIT firefighters, either specialized rescue-air systems or manufacturer designed SCBA shared air systems.

Caution: The SCBA cylinder replacement method should not be used to re-supply air. The process of a firefighter "going off air" long enough to uncouple the high-pressure line connection from the empty cylinder, release and remove the cylinder from the SCBA frame, then replace and re-couple the full-air cylinder is very difficult and dangerous during normal firefighting operations. The SCBA cylinder replacement method during urgent and often unstable conditions during a RIT rescue operation can easily result in failure and cause RIT firefighters also to become victims.

Forcible Exit–Wall Breach

The skill and techniques of "forcible exit–wall breach" requires proper training and the mentality that you can "walk through walls." When normal routes of escape (e.g., hallways, stairways, rear porches, fire escapes) become inaccessible, it might require firefighters to breach a wall, remove studs, and move furniture to enter another room that is safe from fire. They may even have to exit through an exterior wall and escape the building altogether or provide a path for escape due to structural collapse. In such cases, this skill is best performed using a hand tool of some sort for breaching, but, in a last ditch effort, a strong kick to the wall or even the use of an SCBA cylinder might do the job. However it is done, the key is to "never give up" and use all of your options and all of your training to escape the fire.

Fig. 5–16 Firefighter Breaching an Interior Wall for Escape (Hervas)

The forcible exit technique is NOT a foolproof option for the following reasons:

1. The building construction may not allow for a rapid or effective wall breach due to the type of materials used to construct the walls (e.g., a wire-lath-plaster wall or tongue and groove). A firefighter can have greater success in breaching a hollow concrete block than an 80-year old exterior tongue and groove wall.

Fig. 5–17 Pierce the Wall with Hope of Not Hitting a Wall Stud or Another Obstacle (Quickly tear the wall downward, removing wallboard or plaster and lath. Then move over to another stud bay and repeat.) (Hervas)

Fig. 5–18 If Needed, the Firefighter's Hands Can Be Used to Rapidly Pull Away Debris to Clear Out an Escape Hole (Hervas)

Fig. 5–19 Once the wall is Opened, Check for Any Obstacles (Make sure there is a floor on the other side. If necessary, using a handtool, strike the wall stud at its lowest point. This will move the stud enough to provide more room for escape. Send the tool through the hole first.) (Hervas)

2. Construction modifications have been made where one side of the wall might be wallboard, but the other side might be tongue and groove wood paneling.
3. Once the wall is initially breached, electrical wiring, conduit, or plumbing might be positioned in the stud space making it difficult or impossible for a firefighter to pass through.
4. Where the wall is breached will depend on what is on the other side—a safe haven or another dead end. Kitchen cabinets, a refrigerator, or closets might not allow for a suitable escape through a wall.
5. The type of hand tool that the firefighter(s) has is another variable. A Halligan bar is usually the tool of choice along with a sledge hammer or flathead axe.

Hang-Drop Method

There have been many times when firefighters have had to forcibly exit windows to escape fire conditions that have cut off their normal path of escape. Situations where firefighters have become trapped and have had to through a window are:

- Primary search operation on the floor above the fire
- Fighting the fire on the floor above the original fire floor
- Loss of water on the floor above the original fire floor
- Structural collapse of the staircase, floor, or roof
- Sudden ignition of flammable liquids due to arson
- Sudden ignition of a natural gas line

The hang-drop method is an immediate last ditch method intended for a firefighter who is forced by fire to exit an upper-floor window by dropping to the ground. Instead of jumping or straddling over a windowsill and then falling the full distance between the windowsill and the ground, the hang-drop method is designed to reduce the distance of a fall between a windowsill and the ground. The firefighter hangs from the windowsill and then slowly drops to the ground on their feet. For example, say the average length of a firefighter hanging from a windowsill hands-to-feet is 7 ft., and the drop out of the window is 12 ft. The fire-fighter will only have a drop of 5 ft., thereby minimizing any chance for injury. It must be reinforced that this is ONLY a last ditch method of firefighter survival, and it is not to be used in training because it can result in serious injury or even death.

Fig. 5–20 Hang–Drop Method From a Window (Hervas)

Emergency Ladder Escape

The emergency ladder escape is designed for rapid escape out of an upper-floor window onto a ground ladder. One firefighter or a team of firefighters may use it. Unfortunately, there will be times when a firefighter will be cut off from a normal exit and have to resort to a window for escape due to a change in fire conditions and/or structural collapse. For the same last ditch reasons that would cause a firefighter to use the hang-drop method, a firefighter having a ground ladder would use this particular method.

Although there is a perception to the untrained eye that a firefighter exiting a window headfirst onto a ladder is excessively dangerous, however, it is, in fact, safer under pressing conditions than the conventional "step-over" method for various reasons.

The escaping firefighter is staying low to the sill instinctively ducking under the heat and smoke venting out the window. They also have a low center of gravity while exiting the window. While the firefighter is lying face-down on the windowsill to position for escape onto the ladder, there is a reduced risk of losing balance and falling. The alternative of using the conventional step-over method over the windowsill and onto the ladder can result in a loss of balance and a subsequent fall. With the emergency ladder method, one or more firefighters can exit the window much more rapidly, allowing them to escape their dangerous situation safely.

Fig. 5–21 Firefighter Being Forced Out Second Floor Apartment Window (Hervas)

The following was reported using what is known as a Chicago Fire Department Form 2 by Lieutenant Chris Loper. He wrote:

Bureau Operations | **Date** March 22, 2002

Company Truck #32 | **Address** 2358 S. Whipple Street

Subject: Actual use of RIT training at a fire

I respectfully submit that F.F. Dospoy of Truck #32, while in the performance of his duty; particularly searching the second floor of the fire building at 2442 W. 25 Street, at 0242 hrs. for three missing children; had the occasion to exit the second floor window using the technique taught at RIT training. Particularly, exiting the window headfirst, his bringing legs overhead. This was not only helpful, but necessary in order to avoid the extreme heat in the room at the time. Please convey our thanks to the training staff at the Chicago fire academy. We hope this training continues.

When firefighters arrived at a $2^1/_2$-story, wooden frame house with a fast-spreading arson fire on the first floor, a mother of three children was screaming that her three children were trapped on the second floor. The fire had not only extended into the attic but was also spreading to a nearby apartment building as engine companies were struggling with frozen fire hydrants and a consequent delay of sending water. Firefighters entered the second floor from the front initially and then progressed to the rear where the frantic, non-English speaking, mother redirected them. A 16 ft. straight-frame ladder was raised in the narrow gangway, and as firefighters from Truck #32 entered the second floor window, fire from the first floor blew out the window and impinged the ladder.

Part of the fire was temporarily knocked down with a $2^1/_2$-in. hand line from booster tank water, but that soon ran out. All three children were found, but as the first was rescued, the fire had lit up at the ceiling level, forcing two firefighters to exit through two separate windows. Firefighters Dospoy and Herrea had just been trained on the emergency ladder escape and both had used the method almost simultaneously. Firefighter Bob Herrea assigned to Chicago F.D. Truck #32 stated, "Staying low, getting my head out into the cool air, then laying on the ladder, and using my RIT training, saved my life."

Indications for use of the emergency ladder escape

1. **Only during EMERGENCY conditions when a firefighter's life is endangered should this procedure be used on the fireground.** At any other time, the firefighter(s) should exit in the most appropriate manner given the type of ladder, position of the ladder, weather conditions, etc.

2. The emergency ladder escape is a procedure for firefighter survival that is typically performed from windows on the second or third floors.
3. The escaping firefighter should be trained in performing an emergency ladder escape before attempting it. If not trained, then serious injury or even death could result.

Recommendations for using the emergency ladder escape

1. The ladder tip must be set at or just below the windowsill. If the ladder tip rises even an inch above the windowsill, a loss of balance and/or entanglement could result. Exiting firefighters could become hung up on the ladder, slowing or stopping the emergency escape.
2. Ladder manufacturers mandate that firefighting ladders are to be operated at a 75° angle at all times. In the emergency ladder escape, the ladder will generally be used at an approximate 60° angle.
3. The reduced ladder angle will result in a greater likelihood that the ladder may "kick out" if it is NOT properly heeled. Because the ladder tip is not adjusted in height to a sill by the fly section (extension), but by the position of the heel, the ladder will generally have a slightly less angle than 75°.
4. If the ladder angle is steeper than 75°, the escaping firefighters' grip is reduced, and they may become more "top heavy," causing a dangerous imbalance and increasing the risk of a fall. Because less weight can be positioned on the ladder when it is in a reduced angle position, gravity can take over, and a loss of control can result.
5. If the ladder appears too steep to the escaping firefighter, whoever is heeling the ladder must pull the ladder out from the bottom enough to reduce the angle but NOT so much as to cause the ladder tip to drop too far below the sill.
6. The ladder must be heeled (butted) before escaping onto the ladder. During an emergency split-second decision, any available person (firefighter, police officer, or civilian) can heel the ladder. If a ladder is positioned on soft ground, most often it is self-heeled into the ground.
7. If more than one firefighter is exiting, the ladder can be hand-held at the tip in the event the ladder cannot be initially heeled from the ground. Once the first firefighter has reached the ground, the ladder can then be heeled.
8. Use extreme caution when the ladder is wet or frozen. The ladder will become slippery and the reduction in the firefighter's grip can result in a fall.

Procedure for using the emergency ladder escape

1. Set the ladder tip even with or just inches below the windowsill.
2. Thoroughly clean out the window of glass and framework. (Make a window into a door.)
3. The escaping firefighter must slowly, and with control, lean onto the windowsill and initially grasp both ladder beams.
4. The escaping firefighter must HOOK the right arm on Rung #2 with the elbow. Either the right or left arm can be used, but it is recommended regardless of any individual firefighter's dominant arm, that all personnel perform the skill in same manner. Typically, since most firefighters are right handed, hooking the right arm should be procedure for this skill. Hooking the right arm on Rung #2 secures the firefighter to the ladder.

Caution: It is possible for the firefighter to accidentally grasp Rung #2 with an under or over-grip instead of hooking it with the elbow. Grasping Rung #2 will not secure the exiting firefighting to the ladder and can result in a serious fall off the ladder.

5. The escaping firefighter should slide their LEFT HAND down the left beam and use an over-grip on the center of Rung #4. The left

Fig. 5-22 Ladder Tip Positioned at or Below Windowsill (Hervas)

Fig. 5-23 Hook the Right Arm on Rung #2 with the Elbow (Keep the left hand on the beam, staying secure and controlled on the ladder at all times.) (Hervas)

Fig. 5-24 Hooking Rung #2 with the Right Arm as the Left Hand Slides Down the Left Beam to Rung #4 (Hervas)

arm will provide the control and stability needed as the firefighter rotates on the ladder to an upright position. It is best to remember this skill as, "Hook #2, go to #4."

Caution: The exiting firefighter must keep the left hand on the beam until reaching Rung #4 as opposed to blindly reaching for the rung. If the firefighter blindly reaches for the rung and misses, the rotation on the beam will become uncontrolled resulting in a possible injury to the right arm hooked on Rung #2 or even a serious fall off the ladder.

6. As the firefighter is exiting the window, they should bend their knees as their left hand grasps Rung #4 with an overgrip. Then tuck and slide the RIGHT thigh along the left beam, rotate into an upright position, and then set both feet on the closest rung. The right arm will be hooking Rung #2 as the left arm is grasping Rung #4 for control.
7. Execute ladder descent (the method of ladder descent will depend upon whether one or more firefighters are escaping through the window).

Fig. 5–25 The Right Arm Hooked on Rung #2, and the Left Hand Grasping Rung #4 with an Overgrip, "Hook #2, go to #4" (Hervas)

Fig. 5–26 The Rotation on the Left Beam is Slow and Controlled as the Firefighter Rotates to an Upright Position (Hervas)

- If only one firefighter is escaping, it would be appropriate to descend the ladder rung-by-rung.
- If more than one firefighter must use the emergency ladder escape and/or a building collapse is imminent, then once rotated and upright, the exiting firefighters can use an "emergency ladder slide" that will allow multiple firefighters to descend and clear the ladder rapidly.

Emergency ladder slide

The "emergency ladder slide" is designed to be used in the event any firefighter must escape off a ground ladder, whether descending from a roof, window, or in any other operation that requires a firefighter to be on a ladder. Indications for use of the emergency ladder slide include the following:

Fig. 5–27 Once in an Upright Position, Grasp the Underside of Both Ladder Beams (It is important to extend both arms to provide space between the firefighter's chest and the ladder to clear radio microphones, PASS alarms, handlights, etc.) (Hervas)

- When more than one firefighter must use the emergency ladder escape
- During an emergency escape off of a collapsing roof
- In the event of a structural wall collapse
- Due to an unpredicted change in fire conditions (e.g., backdraft)
- A falling ladder or one that is about to fall as a result of soft ground, being struck by an object, or being improperly set

Procedure for the emergency ladder slide

1. Once in an upright position, grasp both of your hands under each of the ladder beams.
2. Position your knees on the outside of ladder beams.
3. Extend your arms to provide enough space between the ladder and your chest to clear PASS alarms, air gauges, hand lights, SCBA facepiece, etc.
4. With control, slowly descend the ladder sliding down the ladder beams. You can maintain control of the slide by gripping with both of your hands and compressing your knees on the ladders beams. This control will eliminate any sudden impact on the ground, which can result in injury.

Contraindications for the ladder slide technique

- The ladder is wet or icy.
- The firefighter(s) is extremely fatigued.
- The firefighter(s) has not been trained on the ladder slide technique.
- The firefighter(s) is injured and physically disabled.

Emergency ladder escape safety system

While learning or practicing the emergency ladder escape skill, it is imperative to use a safety system to ensure that all firefighters are secure with fall protection during training. One way of making the training safer is by using an anchored safety belay or delay line. Realizing that each training facility is different as to available high anchor points for a safety belay line, it may be necessary to consult a trained, technical rescue specialist to establish the most reliable safety system. Mandatory components of the safety system are as follows.

- A "spotting ladder" positioned to the right of the escape ladder
- A secondary system (e.g., cable, anchored block) to secure each ladder in the event the ladder is not heeled properly
- A high anchor point to secure the belay line (e.g., ceiling rafters, anchor bolt, ladder)
- A high anchor point "basket" made of 1-in. rescue webbing or rescue-approved anchor strap
- A Munter hitch or approved mechanical belay device attached to the carabiner and basket
- 10 to 11 mm kernmantle rope to serve as the belay line

Fig. 5–28 Ladder Slide Technique (Hervas)

Fig. 5–29 High Anchor Point Utilizing Ceiling Rafters (Hervas)

Fig. 5–30 High Anchor Point Utilizing a Ladder (Hervas)

Fig. 5–31 Position of Spotting Ladder and Instructor, Belay Line, and Instructor at the Escape Window During a Training Session (Hervas)

- Large carabiner to secure the firefighter to the belay line
- Ladder escape belt(s) meeting NFPA 1983 requirements on fire service life, safety rope, and system components, 1995 edition
- One instructor on the spotting ladder to "hand spot" the escaping firefighter
- One instructor on the spotting ladder to control the belay line from the escape window
- At least one firefighter at the base of the escape ladder at all times to heel the ladder, thus providing safety for the descending firefighter

Recommendations for training

- Explain the indications and recommendations for the emergency ladder escape and the indications and contraindications for the emergency ladder slide.
- Review the ladder placement at or below the windowsill.
- Demonstrate both the approved and unapproved methods.
- Have each firefighter attempt the first skill without their SCBA to master the techniques without the additional SCBA weight and center of balance difficulties. After the first or second attempt, the SCBA should then be introduced as part of the training.
- Ensure that all of the safety systems are in place and that this skill should not be attempted at incidents unless under emergency conditions as a last resort method to save themselves.

Emergency Rappel Methods

Emergency rappel is a last ditch survival method used when a firefighter is forced off an upper floor or roof due to a collapse or change in fire conditions. It is executed when a ground ladder, aerial device, or alternate method of escape (e.g., penthouse staircase, fire escape ladder, or neighboring roof) is not available. Rappelling of any type requires extensive training and attention to safety. Emergency rappelling to escape danger requires even more intense, specific, and continuous training. Emergency rappel has a measure of risk beyond many of the firefighter rescue and survival skills, yet it is better than the alternative of having to jump from an upper story to the ground.

As is the situation with any firefighter rescue and survival skill, some skills are more applicable in some areas of the country than others. It must be realized that emergency rappel methods have not been taught in many fire departments for the following reasons:

1. Many fire departments find it cost prohibitive to provide the proper rescue rope, carabiner, and rope bags for all of their personnel.
2. The logistics to provide refresher emergency rappel training for all trained personnel is difficult.
3. Many towns and cities have access around at least three, if not all four, sides of their buildings that allow for the placement of ground ladders, reducing the need to escape down a rope. It is important to note that firefighter rescue and survival training must emphasize the importance of aggressive laddering around any fire building to the extent possible.
4. If the windowsill is not cleaned of all broken glass or sharp metal window frame pieces, it could cut the rope and thus put the escaping firefighter in grave danger.
5. An emergency escape off a flat roof could be complicated by a lack of solid anchor points. On some buildings, it can be difficult to find anchor points or use structural components due to:
   - Weathered stonework such as corbelling, coping, and parapet walls
   - Corroded metals, sharp gutters, and weakened cables
   - Deteriorated mortar in brickwork
   - Brickwork without bulk and stability (e.g., small chimneys)
   - Vents and stacks made of sheet metal
   - Flashing, gutters, and downspouts that are weak and loose
6. In some cases, a firefighter's lack of agility, age, injury, and other similar factors also need to be considered.

Indications for the use of emergency rappel methods

- The firefighter must have the training and personal rappel equipment to effectively and safely perform this skill.
- The personal rappel equipment would be available because it was stored in a turnout gear pocket or affixed to a safety belt where it would be readily accessible.
- The only remaining option to escape is to jump (hang-drop method) from a window or roof.

Body rappel method

The body rappel method was introduced by the New York City Fire Department. The body rappel method consists of using kernmantle rope of no less than 10 to 11mm in size to encircle the upper torso as it is threaded under each arm. As the firefighter descends, the upper torso, along with a firm grip of the hands, provides enough friction for a controlled rappel toward the ground, balcony, fire escape, secondary roof, or any other surface that provides an area for escape.

Recommended equipment for an emergency rappel consists of:

- Approximately 35 to 50 ft. of (minimum size) 10 to 11mm size kernmantle rope
- One carabiner
- A rope bag

Procedure for the body rappel method

1. Try to "buy" time by closing the door to the room or using any other fixture (e.g., standing up a couch or a mattress, close closet doors, etc.) to block and slow fire extension. Once the window is opened for escape, it must be realized that heat and fire will be drawn into the room very rapidly.
2. Identify a fixed anchor point to tie off the running end of the rope so as to use the carabiner. Such a fixed anchor point could be:
    - Heavy piece of furniture such as a full-size couch, piano, dresser, bed frame, etc.
    - Iron-heating radiator
    - Wall area between two close windows (e.g., wrap webbing or rope around narrow wall)
    - An exposed stud in an opened wall
    - Supports for large machinery
    - Stairwell support beam and column
    - Structural column and projected structural beam
    - Brickwork with large bulk

It is important that the anchor point be as close to the window (thereby as distant from the fire) as possible. For instance, tying off the rappel rope to a piece of furniture close to the hallway door in the path of the advancing heat and fire may not give the firefighter enough time to complete the rappel before the rope burns through.

3. If a fixed anchor point is not possible, the firefighter must resort to a moveable anchor point. This will involve using a firefighting tool such as a Halligan bar, pike pole, or axe that is positioned at a 45° angle in a lower corner of the window. The rope is then wrapped two to three times around the tool and secured with the carabiner. As the firefighter exits the window, constant tension must be kept on the tool to maintain its ability as an anchor point.
4. Once the anchor is set, the firefighter can straddle the window placing the rope overhead and around the shoulders; the rope will at that point rest on top of the SCBA frame. Each arm can then be lifted to secure the rope underneath.
5. When using a movable anchor point, it will be important for the firefighter to lean forward and downward out the window to maintain constant pressure on the tool as it is kept firmly set in the lower corner. By way of a warning, if the tension on the tool is lost, there is a good chance that the tool will fall back into the room and no longer be a safe anchor point.
6. Both the standing and running ends of the rope are then placed together in the outward hand (placing maximum friction on the rope so it will not slide as the firefighter initially exits the window), as the inward hand holds onto the windowsill, thus providing stability and control as the firefighter exits the window.
7. The firefighter then slowly rolls out of the window with full weight and dependency placed on the rappel rope. As the firefighter rolls out, the inward hand releases the windowsill gripping the standing side of the rope.

Fig. 5-32 Haligan Bar Used as a Moveable Anchor Point in the Lower Corner of a Window with a Rappel Rope Tied On (Kolomay)

Fig. 5-33 Exiting Firefighter Straddling the Window Preparing to Descend Using a Life Line (Hervas)

Fig. 5–34 Exiting Firefighter Using the Windowsill to Control the Exit While the Rappel Line is Secured with Constant Tension and a Firm Grip Before Descending (Hervas)

8. As the outward hand slowly releases some grip on the running side of the rope, the rope will slide around the firefighter as a friction device, allowing a controlled descent.

Mechanical rappel device method

Instead of using the firefighter's body as a friction device, the mechanical rappel device method requires the use of a descending device (e.g., Figure 8, carabiner, GriGri ®) that is attached to a rated rescue harness. The fire service sales market has made available different types of emergency rappel systems that are complete with rope, rappel device, and rope bag. Some systems are intrinsic in the turnout coat and pants as well. Whatever the system used, the indications of use for an emergency rappel method are the same, although the procedure may vary slightly depending on the type of system and rappel device used. In any case, the manufacturer's recommendations for use and safety must be observed.

Fig. 5–36 The Firefighter Safely Descends Along the Rope to the Ground (Hervas)

Fig. 5–35 The Firefighter Safely Descends with Control Along the Wall (Hervas)

Training recommendations

1. Review with hands-on training methods the anchoring of the rappel rope to fixed and movable anchor points.
2. Set a spotting ladder and position an instructor on the ladder to both instruct and ensure the safety of the exiting firefighter.
3. Secure a belay system on the exiting firefighter during all training sessions to back up the rappel line.

Components of a Belay System

- A firefighter who is tied to the belay rope and is at risk of falling.
- The belay rope that is attached to the firefighter.
- A harness worn by the firefighter and attached to the belay rope.
- A belay device. The rope is run through the belay device and controlled so that, should the firefighter being belayed fall, the belay device, under the control of the belayer, holds the rope.
- The belayer, who will control the belay device and the belay rope. The belayer's main duty is to brake the rope should the firefighter fall.
- An anchor attached to the belay device. The anchor should be able to hold the highest shock load resulting from a fall of the firefighter being belayed.

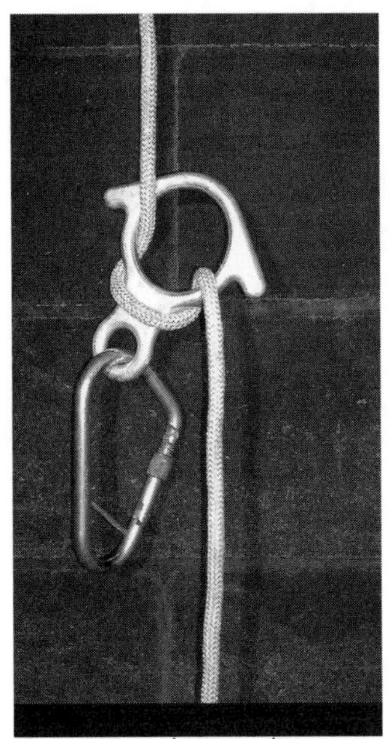

Fig. 5-37 Use of a Descending Device for an Emergency Rappel (Hervas)

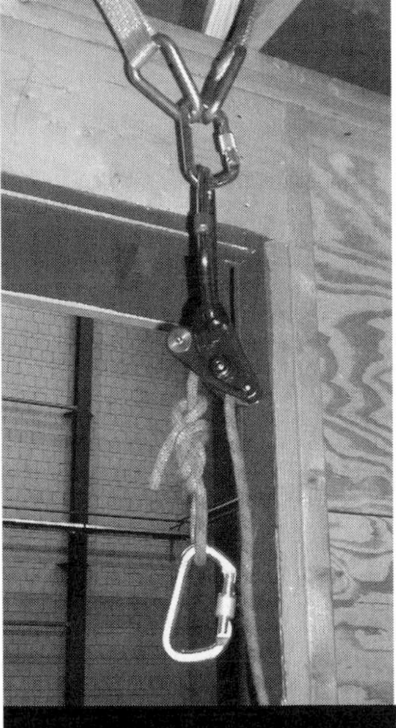

Fig. 5-38 Fixed Anchor Point with a Belay Device (GriGri®) and Rappel Rope Attached (Kolomay)

Fig. 5-39 Fixed Anchor Point with a Munter Hitch, Carabiner, and Rappel Rope Attached (Hervas)

6

Search Operations For Missing and Lost Firefighters

Search operations for missing and lost firefighters assume a very different perspective from search operations for civilians. This is not to imply that there is greater value of one life over another. For that matter, it is acceptable for firefighters to routinely search for civilians who are missing or trapped, but it is not acceptable for firefighters to miss or lose another firefighter. To miss or lose one of our own can lead to a very emotional, unplanned, adrenaline-driven search operation that could not only fail but cost additional rescuer lives.

Lt. Patrick Lynch of the Chicago Fire Department 11 explained the frustration of attempting to search a bowstring-truss roof in an auto tire service center for two missing firefighters who were later found dead on February 11, 1998. Upon the arrival of the rescue squad company to which he was assigned as a firefighter, the company was ordered to search for the victims even though there were heavy fire conditions and no information as to the whereabouts of the missing firefighters. There was no communication with the missing men, and fellow firefighters were too stunned to provide any positive information to help the search operation. It was believed that the missing firefighters had to be back in the service shop. The rescue squad entered through the front door into a small showroom area and then moved further into the building through a second door that separated the showroom from the service area. In spite of their efforts, the rescue was unsuccessful. The first victim was found hours later approximately 12 ft to the right of the showroom door entering the tire service shop, and the second victim was found approximately 45 ft from the same door.

The frustration the members of the rescue squad felt was due, first of all, to a lack of search information. Far more frustrating was that it was later found that the victims' PASS alarms (of the stand-alone version) were not activated and there was no "game plan" for the search operation as the fire conditions were worsening. For Lt. Lynch, all of the these factors, combined with the fact that he personally did not feel prepared enough to execute an aggressive organized search for the missing firefighters, resulted in frustration that has remained with him throughout his life.

It is unknown whether these victims would have lived had a more effective search been possible. What is important to learn from this story is that any responding rescuer or RIT must be trained and adequately supported as much as possible when confronted with a Mayday distress call. Before moving forward into search operations for missing and lost firefighters, it will be important to review key points in basic search operations concerning search plans, tools, accountability, and safety.

Basic Search Operations

The following are the three basic search operations for which RIT members should be trained:

- **Primary search**—a rapid interior search for victims before control of the fire is accomplished. This search only hits the obvious areas and certain target areas (e.g., beds, behind doors, in bathtubs, etc.).
- **Secondary search**—a complete and systematic search of the fire area, the fire floor, the floors above, and any other areas while the fire is being controlled. This search procedure is generally somewhat slower than the primary search.
- **Final search**—a complete and systematic search of the entire structure and the areas surrounding the structure. Adjacent roofs, lightwells, airshafts, yard areas, alleys, basements, and any other area not in the immediate fire area but where victims could possibly be found. This procedure is considered more of a recovery mode than a rescue.

Primary search procedure

1. Search the fire area first to reach those victims in the most immediate danger, that is, those closest to the fire. For the firefighter, the risks must be weighed against the gains that a victim can be saved. Experience, training, and some luck will help in making a prudent decision.

Fig. 6–1 Make Sure to Search Cribs by Hand, Not by Tool (Kolomay)

2. Search the fire floor second, and then work away from the fire area out to surrounding exits (doors and windows), to bedrooms, and any other common living areas.

Fig. 6–2 During VES, Be Sure to Close Doors as Needed to Reduce the Chances of Fire Drafting into Your Search Area (Unknown)

3. Open the rear or alternate doors and/or windows to allow the engine to reach the fire area so that personnel can "open" the rear door(s) and windows to vent, enter, and search (VES). (According to an informal survey of numerous firefighters involved in search operations, approximately 70% of fire victims are found near the rear of fire buildings. This fact has also been confirmed by the State of New York.)[12]
4. Search the floor above next, provided that you have the available personnel. It is optimum if this floor can be searched at the same time as the fire floor. If personnel are not available, then reaching stairwells, windows, and bedrooms are a priority.
5. Vent, enter, and search (VES) for the primary search. (This must be coordinated with the IC or sector officer.) When fire conditions and/or structural conditions will not allow primary search teams to enter conventionally, then it is effective to enter through lower windows or using ground ladders or aerial ladders to

reach search areas. Remember to close the doors to the rooms being searched as needed to reduce the chance of fire drafting into your search area.

Primary search tips

1. Check the exterior window and door layout (most importantly for a way to exit if conditions become dangerous).
2. Check to see where obvious fire and heavy smoke pressure is visible from the exterior.
3. If entering through a window, "Make a window, into a door!" Clear all glass, rails, framework, and drapery before moving on to search.
4. Officers communicate your search plan to your firefighters. Make past training pay off by communicating.
5. Listen for victims—slow or stop SCBA breathing to listen.
6. Know the progress of the engine company. Is fire extending toward the search? Is an additional engine company needed, or is only a hoseline needed to complete the search?
7. When passing or going above an engine company to search, tell the engine officer. If the engine has any difficulty (e.g., burst line or overwhelming fire conditions), then a second line can be positioned and/or the search team can move to a safer position.
8. Always deploy two tools—stage one tool and search with the other. Use tools during a primary or secondary search to:
   - Extend search areas (increase your reach)
   - Ventilate windows
   - Remove a victim
   - Breach walls
   - Sound floors and stairs for holes (lightweight construction)
   - Chock doors
   - Indicate your presence in a room or on a floor
9. Stay on a wall during moderate or heavy smoke conditions. To extend out from the wall to the center of the room and cover more ground in your search, posi-

tion a tool (e.g., Halligan bar) against the baseboard, and extend your reach to sweep the room with your arm and hand to identify bodies. This method allows you never to lose contact with the perimeter wall.

10. Move objects only enough to perform the search. Do not throw furniture and other objects wildly. This may possibly result in blocking and confusing your path and perhaps even cover a victim.
11. Check behind all doors, and chock or block all doors. If both searchers must enter a room, then place a pike pole or bar across the threshold to prevent the door from closing fully, as well as to help indicate the presence of firefighters searching in that particular room.
12. If feasible, ventilate windows during the primary search. Know the building construction, void spaces, and occupancy. Know the heat conditions around you, and know that conditions can violently change when "opening up" as it could possibly draw fire in your direction. Make sure a working hand line(s) is confining the fire. Remember that premature venting could endanger those who are trapped as well as the rescuers.
13. Check unusual areas during the secondary search. (These are some examples of where victims have been found. There are many others to list.)

    - Bathtubs and showers
    - Piles of clothes in closets
    - Kitchen pantries
    - Behind drapes
    - Large dresser drawers

14. Inform the IC of search progress and fire conditions. In many cases, such progress reports will shape the fire strategy being used. Progress reports will reflect things such as the need for roof ventilation, the need for additional ground ladders to be positioned and raised, the need for additional search teams, and/or the need for hoselines to be deployed.

Personal search rope operations

Not unlike the thermal imaging camera, the personal search rope is another tool for safety. A personal search rope can be defined as a rope carried by an individual firefighter to conduct team search operations. The primary question any firefighter should ask when deciding to use a personal rope during a search is "Why?" What is the determining factor that calls for use of a personal search rope? If ever in doubt about whether to use a search rope before committing to a search, lean on the side of safety, and deploy the rope. A search rope should not necessarily allow you to get in farther; instead it should allow you to get in and out more safely.

An experience that underscores this advice occurred at a rear entry door of an unprotected, non-combustible, construction-strip warehouse that measured 200-ft X 60-ft X 20-ft. A squad company was ordered to conduct a search of the rear section of the warehouse area next to where the fire was found. They found a light heat condition and a heavy smoke condition about 5 ft. above the floor. The officer first thought of using a search rope but resorted to instructing the company to stay on the right wall with the thought that the smoke conditions would not worsen given the height and size of the building. Unfortunately, once committed about 40 feet inside, the officer found that the smoke was dropping quickly and the ability to search away from the wall was limited. Additionally, one of the firefighters had almost fallen into a deep truck well. It was at this point the officer knew it was a bad decision not to use a search rope even though at first the conditions did not look life threatening.

Search rope rules. The following is a list of basic search rope rules as well as the reasons why the search rope is a valuable tool in performing rescue operations.

1. A firefighter with a portable radio and large hand-held flashlight should always be positioned at the entry point to the search area. A firefighter positioned at this point of the search area will be able to account for the number and identity of the firefighters searching, assist with communications with the search team via portable radio or verbally, and assure the entry point remains safe from deteriorating fire conditions, collapse, or even the simple act of the entry point door closing accidentally. The positioning of a large hand-held flashlight, as opposed to a small personal helmet light, in the threshold of the entry point will serve as a reference for the search team.
2. During rapid intervention operations, a search rope should always be deployed, especially when searching with a thermal imaging camera in a wide area such as a factory, office area, or commercial building. (A search rope is usually not needed in single-family dwellings due to the availability of walls, windows, and doors for reference and escape.) Because firefighters are entering an unstable incident, it is important that the search rope not only serve as a safety tool for escape purposes but also provide back-up companies with direct guidance to locate the RIT if they need any type of assistance.
3. Always tag the anchor end and secure the running end of the search rope. For accountability reasons, it is important that the anchor end of the rope have a durable identification tag secured to it with the identity of the fire department

or company. Equally important is to secure the running end of the rope to the rope bag and/or pocket that the rope is deploying from. Both of the authors have experienced incidents where the search rope was not secured, had come to its end, and then simply pulled out of the pocket. Unknowingly, we no longer had the safety of a search rope.

4. Don't carry enough personal rope to "hang yourself." Personal ropes measuring 35 ft to 50 ft in length and 8 to 10 mm ($^3/_8$ in.) in thickness are most common (Some firefighters have been seen pulling a wad of rope that measures 200 ft. out of their turnout coat pockets only to be unable to find the beginning or the end of the rope.) Designated search rope bags, as opposed to personal search rope, are commonly set up with 100 ft. to 200 ft. of rope, $^1/_2$ in. to $^5/_8$ in. thick, and are meant for more complicated search operations such as wide-area search and rapid intervention incidents. One very important point to consider is that the search rope is a safety device, not only to provide a direction out but also to prevent the search team from over-committing in heavy smoke conditions. Given high ceilings and the inability to recognize dangerous heat levels as a result, a search team can advance 30 to 40 feet into a building within seconds. Without a search rope, the search team will be dependent on landmarks, walls, and points of light, which can all disappear in seconds.

Most experienced firefighters carry no more than 50 ft of search rope because the limited distance allowed by this length will allow them to exit as quickly as they entered, thus preventing the firefighter from reaching a point of no return. This helps to prevent firefighters from falling victim to the following scenarios:

- Rapidly deteriorating fire conditions causing the firefighters to be caught in a flashover
- Not having enough SCBA air to make it back to the entry point or any other point of refuge for air
- Structural failure resulting in collapse
- Falling contents such as high rack storage, stacked paper bails, and machinery

Fig. 6–3 Setting Up for a Wide-Area Search Operation with Heavy Smoke Conditions

Suppose you are 40 feet into the building during moderate to heavy smoke conditions. If your search rope runs out at this point, it will help stop you from over-committing to the dangerous aforementioned "point-of-no-return." Some firefighters use certain "tricks" such as counting steps, "crawls," or noting landmarks to ensure a safe exit when crawling into a heavy smoke and heat condition. While these strategies may be somewhat useful, the fact is that experienced firefighters are preoccupied with sounding the floor, using the radio, feeling for heat change, listening for fire or victims, and much more. Because of performing these necessary activities as they move in and conduct a search, they often cannot concentrate on counting. The safer, easier, and most effective tool to use to prevent over-committing is the use of a personal rope.

Setting up a personal rope. Setting up a personal rope can be easy, but there are several recommendations that should be followed.

- Use a rope bag rather than storing the rope directly in a turnout gear pocket. The rope bag will not only help protect your rope, it will provide a loop to secure the running end of the rope thereby reducing the tendency for the rope to "knot up" when being deployed. Rope bags for personal search ropes are readily available from numerous distributors and are now designed in various shapes and sizes to conform to the shape of turnout gear pockets whether they are round or square.
- Use kernmantle rope. Although more expensive, kernmantle rope has the thickness (10 mm or $^{3}/_{8}$ in.), durability, and strength to be dependable in even the worst situations. Up to 50 ft. of rope can be easily stuffed into a rope bag without excessive weight or bulk. Again, do not forget to tie the running end of the rope to the bag before loading the bag with the rope.
- Set up the anchor end properly. On the anchor end of the rope, it will be important to tie a figure 8 knot on a bight and affix a spring clip or carabiner to it. This will allow the search rope to be wrapped around an anchor point and secured as the carabiner is affixed back onto the anchor end of the rope.

Fig. 6–4 Various Types of Search Rope Bags (Hervas)

- Affix an identification tag on the anchor end of the search rope. This will provide increased accountability in the event the firefighter positioned at the entry point must enter the building to assist the search team.
- Make the rope easily accessible. If you are right handed, it is advisable to position your rope on the right side of your turnout gear in an easily accessible pocket.
- Store the rope bag in a turnout gear pocket. The less rope hanging out, the less chance there is that it will get hung up. Rope bags have been seen hanging from utility belts and SCBA harnesses in positions that have easily become entangled or wedged on anything and everything. Keep your personal rope secure and protected in a turnout gear pocket.

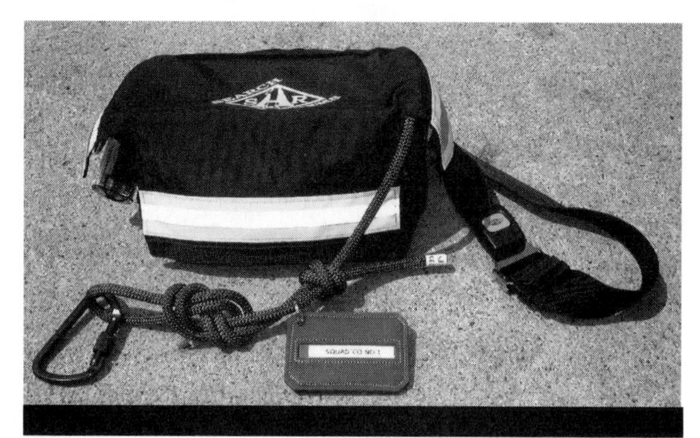

Fig. 6–5 The Anchor End of the Search Rope with a Figure 8 Knot, Carabiner, and Identification Tag (Hervas)

The deployment of a personal search rope is to be done with full personal protective equipment and SCBA by:

1. Reaching into the pocket and picking out the carabiner
2. Tying off the rope onto an anchor point
3. Moving into the building and letting the rope automatically deploy out of the rope bag

If the rope ends, the rope bag will give a hard tug to the binding in the pocket, indicating to the firefighter that their limit has been reached. If there is a chance that the rope bag is too loose and will be pulled out of the pocket, it should be affixed to the pocket.

Thermal Imaging Camera

The thermal imaging camera is an excellent tool, not only for civilian search operations but also for rapid intervention operations. Using a thermal imaging camera can greatly enhance RIT search operations for missing and lost firefighters. Although the thermal imaging camera is becoming a more readily available tool throughout the fire service, it is still cost prohibitive in many fire departments throughout the country. The cost averages between $14,000 and $22,000 per camera. Even if a fire department can obtain a thermal imaging camera, firefighters should be aware that a thermal imaging camera is not without its share of possible problems. The following is a list of reasons the camera may become unusable.

- Technical failure with camera parts
- Physical breakage from a drop or impact
- Loss of power
- Loss of the whole camera from a fall, slip, or from momentarily setting the camera down in heavy debris or smoke conditions

Firefighters, therefore, must not place a high dependency on it or disregard the basics of search because "they can see" through the smoke. They must still use proven search techniques, acquired skills, knowledge, and experience. Observe the following recommendations to best coordinate the use of a thermal imaging camera in a RIT search operation.

1. The RIT officer leading the team should have the camera in order to determine:

   - The interior layout of walls, furniture, aisles, hallways, etc.
   - Structural failures such as holes in the floor, collapse areas, fallen debris, etc.
   - Hot spots where fire can be located in relation to the search operation
   - The immediate results of the scan of the search area for victims
   - The locations of RIT firefighters conducting room searches

2. As the RIT officer advances with the thermal imaging camera, the RIT firefighter should deploy the search rope and stay in physical contact with the RIT officer. If the RIT officer or any firefighter is not trained or experienced in the use of a thermal imaging camera, they will tend to advance much more quickly and thereby separate from the team. Even when trained, during the "heat of the

moment" it can be easy to forget that the rest of the search team cannot see and will not be as likely to move as quickly as the one who can "see" with the thermal imaging camera.

3. As the RIT advances into the search area, the officer can pass the camera to another RIT firefighter so that they can also scan the area and have a mental image of the interior layout before expanding the search operation.
4. If the RIT becomes committed to a rescue operation, the thermal imaging camera will help in visualizing just what needs to be done in terms of disentanglement, knots, clearing of debris, and much more to effect the rescue. One of the problems with RIT rescue operations is that the officer sometimes becomes too physically involved in the rescue operation. Inherently, if the officer holds onto and uses the camera properly, it will somewhat prohibit any hands-on involvement and encourage the officer's role in supervision, communication, and safety.

Fig. 6-6 RIT Search Team Advances with Thermal Imaging Camera (Hervas)

5. If more than one thermal imaging camera is available, the second camera should be assigned to the RIT sector officer to allow him or her to see the interior operations and conditions.

Wide-Area Search Operations

Incident command procedures for missing firefighters

A missing firefighter can be defined best as a firefighter who is reported missing, cannot be located, or cannot be contacted via radio after repeated attempts. The search for a missing firefighter requires an information-intense operation. The greatest risk with a search operation for a missing firefighter is to commit a RIT to a high-risk search when the firefighter may not even be in the building. However, it must also be considered that a missing firefighter could be lost and trapped if they cannot be accounted for and they:

- Were not radio equipped and could not report the Mayday
- Had a radio that was disabled due to water or impact
- Lost the radio from a pocket or harness during a fall or collapse
- Could not operate the radio due to injury from impact or burns

Once the firefighter has been reported missing, the IC must commit immediately to an action plan as follows:

1. Confirm the report and stop all non-essential radio transmissions.
2. Have fireground companies switch to a secondary radio channel (if possible). The primary (or original) radio channel will be reserved for the IC, RIT sector officer, RIT, and possibly the missing firefighter. The secondary channel will be used for the fireground operations, eliminating communications interference with the rapid intervention operation.
3. Obtain any possible information known about the missing firefighter:

    - Firefighter's company and/or name
    - Last known location
    - Fireground assignment (e.g., roof ventilation, second floor search, etc.)
    - Whether the missing firefighter was radio-equipped or not
    - Who the missing firefighter was with last and if any other firefighters are missing

4. Advise all companies that there is a missing firefighter, and identify who it is.
5. Assign a RIT sector officer to the rapid intervention and search operations, and activate the rapid intervention operation.
6. Start a fireground personnel accountability report of all companies.
7. Assign additional companies for additional rapid intervention operations and deploy them as needed to back up the original RIT or for additional search operations.
8. The IC must maintain tactical control of the fire building in the offensive mode if at all possible; this may require a call for an additional response alarm.

Incident command procedures for lost firefighters

A lost firefighter can be best defined as a firefighter who personally reports being lost in a fire building. This report typically is received via radio. One of the main missions of a RIT is to respond to a Mayday distress call from a lost firefighter and to be prepared with personnel, tools, and a plan of action for rescue. Once the firefighter(s) has issued a Mayday distress call, the IC must immediately commit to an action plan as follows:

1. Confirm the Mayday report, and stop all non-essential radio transmissions.
2. Have fireground companies switch to a secondary radio channel (if possible).
3. Establish radio contact with the lost firefighter (if possible).
4. Identify the firefighter(s).
5. Identify the lost firefighter's location—what floor, front or rear, type of room, near a window, identifiable sounds (e.g., air horns, surrounding radios) and lighting (e.g., exit lights, building emergency lighting, fire alarm strobes), etc.
6. Identify the condition of the lost firefighter(s)—low SCBA air, injured, entangled or trapped.
7. Ascertain surrounding fire and structural conditions.
8. For the benefit of any firefighters immediately working nearby, request the lost firefighter(s) to produce repeated sounds (e.g., striking a floor, a metal banister or railing, breaking a window, etc.)

   Caution: With reference to breaking out windows, it must be remembered that in some cases this act can worsen the fire conditions in the immediate area of the Mayday situation.
9. For the benefit of any firefighters immediately working nearby, request the lost firefighter(s) to activate the PASS alarm for a 5-second interval. If there is more than one lost firefighter, only one should activate an alarm to eliminate confusion and excessive sound.
10. Assign a RIT commander to the rapid intervention and search operation, and activate the rapid intervention operation.
11. Start a fireground personnel accountability report of companies.
12. Assign additional companies for additional rapid intervention operations, and deploy them as needed to back up the original RIT or for additional search and rescue operations.
13. The IC must maintain an offensive tactical control of the fire building and call for an additional response alarm.

The importance of information gathering

Information gathering in this operation is imperative and will determine the type and size of a RIT search operation and the risks to be taken. The amount of information from the lost firefighter(s) or about the missing firefighter(s) will determine the size of the search operation. The less information, the larger the search operation. Another factor in determining the size of the search is whether or not there is more than one missing and lost firefighter and if they are separated or not. The size of the building with respect to the floor area, number of floors, and interior configuration (e.g., elevators, high-rack storage, movable office walls) will determine the type of search operation. Fire conditions and structural conditions will determine just how far any RIT will be able to search. A structural collapse will dictate where a RIT can and cannot search based on structural integrity, available void spaces, and possible areas for tunneling.

In 1999, during a RIT search operation in a one-story 500 ft X 600 ft paper warehouse building equipped with a sprinkler system, the chief officer became disoriented and eventually ran out of SCBA air. Numerous hand lines and a master stream had been fighting the fire that was spreading throughout stored paper bales when the victim ordered personnel to evacuate the building due to a sudden change in smoke conditions. An evacuation radio message and apparatus air horns were sounded. However, not all firefighters in the building heard the evacuation signals. Contributing to the initial evacuation confusion, five firefighters had become low on SCBA air, disoriented, and had to utilize buddy-breathing methods for escape. At the same time the victim had also become disoriented and separated from other firefighters and hoselines. The IC then received a call for help via radio from an unknown firefighter (the victim) still inside the building after all other personnel had been evacuated. The IC asked who needed help, and the victim responded by saying "106." The victim was unable to tell the IC his location inside the structure. The IC told him they were sending in the initial RITs, Hazmat #71 as RIT #1 and Heavy Rescue #1 as RIT #2, to find the victim. The IC ordered that all radio traffic be held to a minimum and for all companies to switch back to a different radio channel. Car 100 (the Fire Chief) arrived and was given an update of the situation, including concerns about the structure. Car 100 and the IC concurred that the search activities would remain the priority and ordered both RITs to enter the warehouse and search for the victim.

On the south side of the building, there were 13 overhead dock doors and one standard swinging door. RIT #2 entered the structure with a rope through one of the dock doors. After donning their equipment, RIT #1 entered with a rope, behind RIT #2. RIT #2 went toward the area where the victim was last seen and searched until three of the firefighters' low-air alarms sounded and they exited the structure. The Captain and a firefighter of Heavy Rescue 1 continued the search until their low-air alarms sounded and they then exited the building. With one-hour SCBA air cylinders, RIT #1 searched an area in the front of the structure until their low-air alarms sounded and they also exited the structure. At least 17 additional firefighters reported that they entered the structure to search for the victim at various times during this interval.

Additional companies formed two more RITs (#3 and #4) and entered the warehouse with additional ropes to search for the victim. Throughout the search, the victim radioed that he thought he was in the same location where he was when the smoke banked down. Within minutes, the victim said that he was out of air and was breathing off of the floor and asked if all other personnel were accounted for. The IC noticed that his voice was labored and garbled. The IC asked the victim if he could manually activate his PASS device, but received no response. No further communications with the victim were ever received. The IC then ordered dispatch to notify a mutual-aid department to respond with a thermal imaging camera.

The power to the building was restored providing interior lighting, and RIT #2 reported to the IC that they were changing their air bottles and would be ready to go back inside. The IC told RIT #2 to brief one of the other RITs (RIT #3 or #4) about the locations they had searched and then send one of them inside. RIT #2 then radioed the IC that they had a good idea of the area that they had searched and they were going to re-enter. RIT #2, along with two captains and four firefighters as well as a firefighter from a mutual-aid department with a thermal imaging camera, combined to form a new RIT. RIT #5 entered the structure to continue the search. Following the ropes back to the area they had just searched, a captain from RIT #5 veered off to his right and found the victim. The victim was unconscious, with no helmet, radio, or SCBA. The Captain immediately yelled to the other firefighters for assistance in removing the victim from the building. Unable to locate the victim's pulse, the Captain immediately began cardiopulmonary resuscitation (CPR) and continued it until the other firefighters arrived. A radio call was made to the IC informing him that they had located the victim and were removing him. Firefighters later stated that the smoke was clearing and the visibility was improved. The victim's helmet was found on the floor in an area where he was operating throughout the efforts to fight the fire. His SCBA was found approximately 10 feet away from him. He was found equipped with a PASS device, but it was not turned on. The medical examiner listed the cause of death as asphyxia with carbon monoxide inhalation. The victim's carbon monoxide level was listed at 51%.[12] This case study demonstrates the difficulty of searching a warehouse measuring 300,000 square feet in heavy, cold smoke conditions for one victim who was lost and could not provide accurate information concerning his location.

RITs must be trained in various search operations so they can adjust to whatever situation they may be confronted with when a firefighter(s) is missing or lost. While the IC and RIT sector officer must gain or maintain control of the incident, communications, and accountability, it will be the responsibility of the RIT to follow through with the proper procedures for a search operation for a missing or lost firefighter(s). These procedures include:

1. Report to the IC for information and orders.
2. Determine how many backup and support RITs will be needed.
3. Determine if one or more engine companies will be needed for water protection.
4. Locate an anchor point for the search rope, and activate the thermal imaging camera (if available).
5. Determine the need for an interior reconnaissance before starting a search operation.
6. Start the search operation. The RIT officer must lead the team, providing direction and determining the level of risk. RIT firefighters should only be deployed when in contact with walls or the search rope.

Locate missing and lost firefighters by following these tips:

- Follow hoselines into areas where the victims may be.
- Account for abandoned tools that may have belonged to the victims.
- Look for a hand-held flashlight beam in a smoke-filled room or debris pile.
- Listen for audible alarms from a PASS device or SCBA low-air alarm or, if radio equipped, transmissions from the victim's radio.
- Listen for calls of help, coughing, banging, or any deliberate noise that a victim might make.

Types of searches

The RIT should be familiar with various types of searches that may need to be conducted. Although some aspects of each type of search remain the same, others vary greatly. The RIT needs to be familiar with the procedures for each of the following types of searches.

Single-entry point search. The RIT single-entry point search is designed for a basic rapid deployment from one point of entry into an area where the victim(s) is highly suspected or known to be located. The objectives of this search operation are to conduct a rapid and basic primary search, to provide reconnaissance information on the search and rescue efforts, and to assist in performing a rapid rescue of a missing or lost firefighter(s).

A single-entry point search would be most commonly used in:

- Single-family dwellings, town homes, apartments, and small commercial buildings
- Structural collapse where very few voids exist and/or very few entry points exist
- Confined-space collapse incidents where only one search operation can be conducted at a time due to the risk of a secondary collapse if additional search teams operate
- A structure that has few doors and windows or high security, such as banks and financial exchange facilities that are insulated with steel and fiberglass walls, floors, and roofs

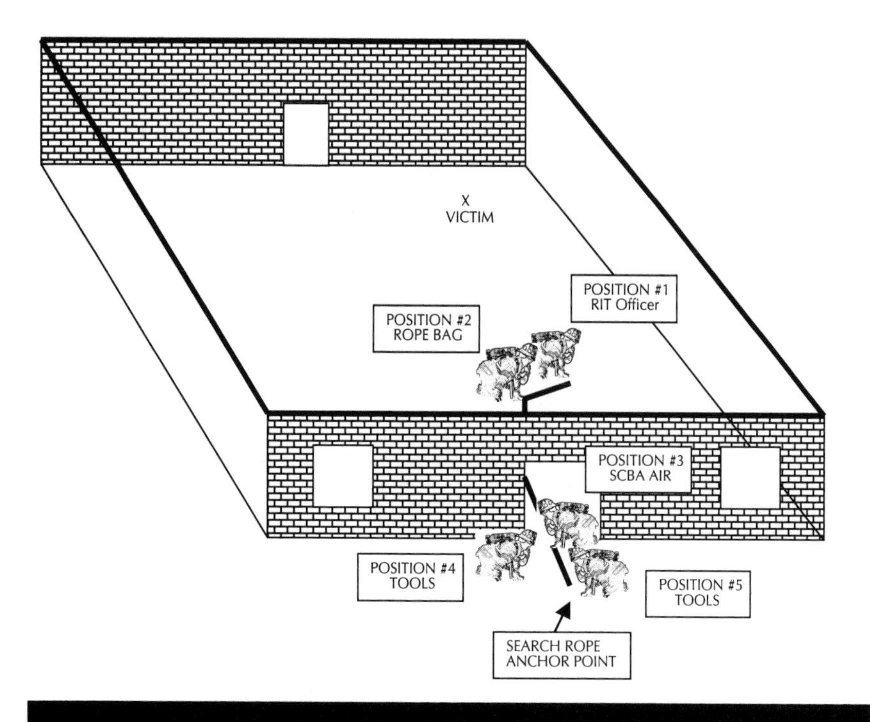

Fig. 6–1 RIT Single Entry Point Search

To execute a single-entry point search, the RIT will:

1. Use any available or specific information in locating the lost firefighter(s). Information regarding where the firefighter victim(s) are possibly located will often determine the best point of entry.
2. Conduct reconnaissance. From the entry-level position, the RIT officer will size up and evaluate the ability to enter along with the risk level. Such reconnaissance may reveal that the closest point of entry to the victim may be too dangerous (e.g., eminent collapse, heavy fire conditions).
3. Deploy a search line by securing one end at the entry point, and advance the rope bag inside the building as the search progresses. Make sure to secure the search rope to a solid and obvious anchor point such as a door frame, door handle, guard rail or post, etc. Remember to keep the rope taut so it will be easier to follow. It will reduce the chances of hanging up while searching, and it will use the maximum length of the rope.
4. If available, utilize a thermal imaging camera to help coordinate the search and help locate the victim(s).
5. Request a hoseline for protection if needed in the search operation and rescue site.
6. If the RIT progress several feet into the building but cannot continue inward due to a burned-out area of floor, such information will cause a second RIT to search from a second point of entry. The search then becomes a "multiple-entry point" search.

Multiple-entry point search. The RIT multiple-entry point search operation is designed to "divide and conquer" areas that are rather difficult to search. When a building or a building's section becomes divided into sectors, it will be up to at least one RIT sector officer to oversee the search pattern. A RIT commander oversees this operation from the command post. This search operation is also basic, rapid, and effective, whether victim information is available or not. A multiple-entry point search would be used in the following situations:

1. Below-grade search operations in a basement, sub-basement, cellar, or below-grade subway station where two points of entry through windows, walk-out doors, sidewalk access doors, entry tunnels, and secondary stairways could be used as entry points.

2. Large-area search operations in residential and office high rises, commercial, and industrial buildings where multiple doors, dock doors, and stairwells could be used for points of entry.

3. Large collapse incidents (e.g., in single-family, multi-family, commercial, industrial buildings where structural components have failed due to fire damage, water load, overhaul, etc.) where several points of entry could be used.

The objectives of this search operation are for the two or more RITs cover more search area quickly and to reach the victim(s) from an easier vantage point. Although the lost firefighter(s) may have originally entered through the front door of a burning store, the first RIT may enter there, while a second RIT would enter a rear door, window, or even breach a wall for a second point of entry. The victim(s) may be very close to the rear and immediately found and rescued out the rear at the opposite end from where they entered.

If the first RIT is having problems with one point of entry due to fire conditions, structural collapse problems, high security systems, or dead-end hallways and staircases, there will be a second point of entry being created simultaneously by the other RIT. In some cases, such alternate search entry points may not be routine. For instance, some examples of alternate entry points may include breaching a wall, opening up heavy security systems, removing trucks from truck docks, or entering from an exposure building.

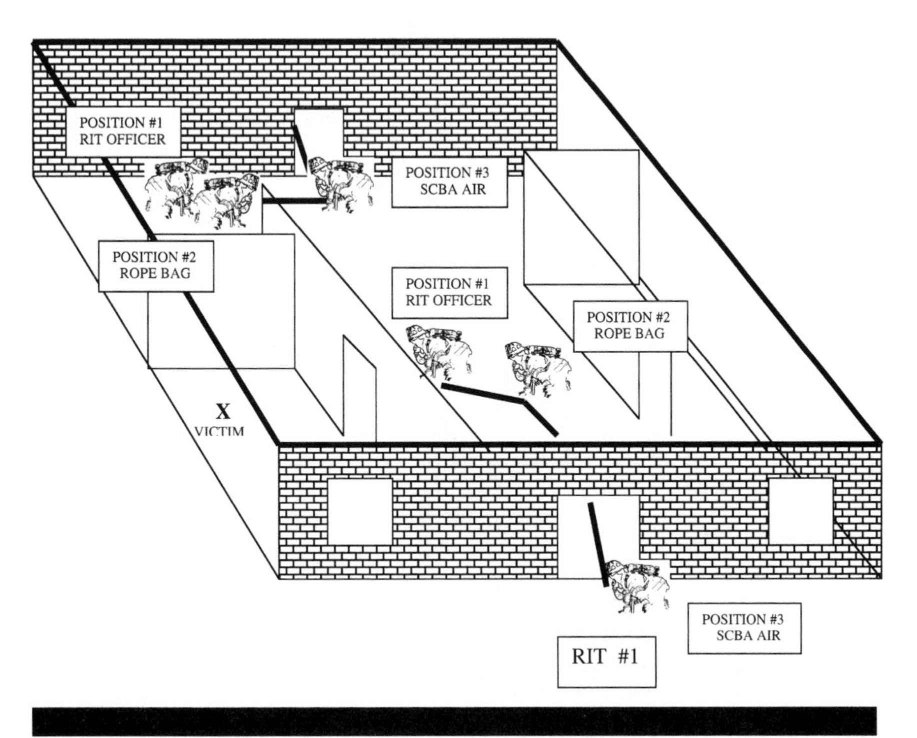

Fig. 6–2 Multiple-Entry Point Search

To execute a multiple-entry point search, the RITs will:

1. Use any available or specific information in locating the lost firefighter(s).
2. Deploy search lines by securing one end at each entry point, and advance the rope bags inside the building as each search progresses.
3. If available, utilize a thermal imaging camera to help coordinate each search operation and help locate the victim(s).
4. Request hoselines for protection if needed for the search operation and rescue site.

Wide-area search. RIT wide-area search operations involve the use of a coordinated search effort with the use of search ropes, a minimum of four firefighters, portable radios, and hand tools (including a thermal imaging camera if available). RIT wide-area search operations may be applied in buildings such as:

- Supermarkets
- Schools and gymnasiums
- Auto-repair garages and bowling alleys
- Auditoriums, theaters, stadiums, and arenas
- Shopping malls and super stores
- High-rise office buildings
- Industrial, warehouse, and cold storage buildings

Fig. 6–9 Typical 1900's Constructed, Cold Storage Warehouse with 18-in. Cork Insulated Walls and Mill-Constructed Floors and Roof (Kolomay)

Not only must a large area be searched, but also maze-like aisles and complicated layouts may have to be navigated. Because RIT wide-area search operations are labor intensive and can become complicated when compared to a single-entry point search, they require comprehensive training that is understood by all RIT firefighters. Wide-area search operations can commit large numbers of firefighters during a rescue effort, increasing the chances of additional problems. It is essential that any wide-area search operation be kept as simplified as possible by using a "divide and conquer" strategy of sectoring the building and then searching the most probable sector(s) where the victim(s) might be located. As much as any fire department may train on a procedure, experience has shown repeatedly that training skills must be adapted and changed to conform to the type of building and current fire conditions. It is also very important for the IC to attempt to improve the fire conditions by confining the fire and ventilating during a RIT wide-area search. Then the RIT will only have to depend on the search rope for safety reasons in the event that smoke conditions worsen, a collapse situation is eminent, a member of the RIT falls victim to injury or a medical problem and must be evacuated quickly, or if members must retreat for SCBA air. There have been incidents when firefighters have been reported missing or lost in cold smoke and the building was then evacuated on orders of the IC. The firefighting efforts were thus hampered because the focus became the rescue efforts (e.g., RIT wide-area search, communicating with the victim, accountability, etc.). Normal firefighting operations were forgotten. Past fatality incidents investigated by NIOSH document lessons that were learned in such operations when roof ventilation had not been completed after the Mayday was called. Had the smoke conditions improved and slightly lifted from the floor, the chances for victim survival would have increased, and the RIT's search efforts possibly would have been more successful.

Positioning hoselines to confine the fire is equally as important as ventilation to allow the RIT wide-area search to be as successful as possible. Hoselines that are operating after the Mayday distress call hopefully will improve fire conditions by reducing the heat, reducing risk of collapse, and in some cases, allowing an operating sprinkler system to be shut down. In cold smoke incidents, open sprinkler heads positioned on high ceilings will moisturize and cool warm or hot rising smoke, causing a heavy blanket of smoke to drop to the floor. Ventilating this type of smoke is very difficult and, at times, impossible until the sprinklers are shut down.

Fig. 6-10 Modern Cold Storage Warehouse Housing 3 Separate 10,000 Square Foot Cold Storage Buildings (Kolomay)

It is important to emphasize again, that during wide area and high-rise building fires, the importance of aggressive firefighting operations during the search and rescue of a missing, lost, or trapped firefighter are essential. Without aggressive firefighting operations during a Mayday incident, the RIT(s) extend themselves into a wide area with an unnecessarily high level of risk. The lack of firefighting operations also reduces the odds that the firefighter victim(s) will survive until a rescuer reaches them.

Wide-area search communication. Given the high level of risk of committing a RIT into a large building that will require the use of search rope to locate a missing firefighter, each RIT member should have a radio. Ideally, the radio channel should be dedicated to the search operation, but a dedicated channel used for the overall RIT operation will work too. The objective is not to have fireground messages jamming up the radio and negating their use.

Verbal commands will be the responsibility of each local fire department. Realizing that there will be variations, such as sector designations, team designations, etc., it is very important to keep the search commands very basic. "Forward," "Return," "Stop," "Turn right," and "Emergency" are examples of clear and concise communications that can be transmitted via radio or face-to-face. Experience has shown that verbal communication is greatly preferred to the use of rope tugs such as the OATH system (1 pull = OK, 2 pulls = Advance line, 3 pulls = Take up slack, and 4 pulls = Help). Searchers have misinterpreted not only the number of tugs but also what they mean. There have also been instances when the search rope has been pulled around an object or the searcher was accidentally kneeling on the rope thus preventing the tugs from transmitting to the searcher who needed to be contacted. It is important that each RIT member receive the necessary training so as to know and use the proper communication while executing wide-area search procedures.

Basic RIT wide-area search ropes. It is necessary for any RIT that is deployed to conduct a wide-area search operation to use a main (wide area) search rope. This is true also when using a thermal imaging camera to conduct searches. There have been many thoughts on the type, size, and length of ropes to be used during wide-area searches. The authors of this book recommend kernmantle rope because of its strength, durability, and availability to the fire service.

Main wide-area search rope. The diameter of the rope can vary, but the objective is to use a rope that is thick enough to hold on to and locate easily with a gloved hand. Therefore, an average rope size of $^{7}/_{16}$ in. or 11 mm is recommended. There are some kernmantle ropes with reflective material woven in the mantle that can help the search effort, but those can be cost prohibitive to some fire departments. The recommended length is 200 ft. The first 50 ft. of rope is reserved for the distance between an exterior anchor point and the starting search point. This first 50-foot section can be indicated by tying a small and basic overhand knot at the 50-foot mark. One of the complications with inserting large knots (e.g., butterfly or figure-8 knots) is that they hang up in the rope bag not allowing the rope to deploy smoothly. Some experiences have shown where a large knot had lodged in the rope bag so tightly that when it was pulled, the whole bag of rope dumped onto the floor. It is, again, important to keep any indicating knots small and the number of knots minimal. The 50 ft leader of main search rope can be easily used up if it is anchored onto a stationary object outside the building and then led inside the building (e.g., anchor to a crash post outside of a truck dock, then enter the factory through a receiving area) to reach the starting search point. This will leave a manageable 100 ft to 150 ft of main search rope for the search operation in the event rescuers must exit the building in a timely manner due to low air, a weakened structure, or a change in fire conditions. If the rescuers need more main search rope to extend the search, the RIT chief and/or IC must be alerted to just how far the rescuers should be allowed into the building. At this point, consideration must be given to having additional search teams enter from alternate entry points, which might be closer to the expanded search area.

Secondary wide-area search rope. This rope will generally be smaller in diameter and used to search off the main search rope. An average size of $^3/_8$ in. or 10 mm is recommended. The length should be no more than 50 ft, which will allow for some distance and mobility but not too much as the rescuer extends off from the main search rope. The secondary rope is best constructed of kernmantle rope.

To execute a wide-area search, the RITs will:

1. Locate an entry point (e.g., dock door, man door). Not unlike a RIT multiple-entry search, all possible entry points need to be identified and several RIT wide-area search operations may have to be deployed. It is important to chock any door being used as an entry point, especially a roll-down, overhead dock door due to overall failure of the door and track when subjected to high heat. Using a ladder to block the door from underneath or affixing vice-grip pliers onto the door track to block the door rollers are common methods used to chock an overhead door.

Fig. 6–11 Overhead Rolldown Door with Heavy Smoke and Heat Conditions (Hervas)

2. Determine a need for a hoseline to accompany the search operation. Size-up of fire, heat, and smoke conditions will determine the need for a hoseline, not only for extinguishment while advancing but also for safety reasons. As the hoseline advances with the search rope, it can also serve as an alternate guide in and out of search area.
3. Set up RIT wide-area search at the entry point. The following points must be accomplished before committing searchers to the interior:

- Obtain most recent "victim information" (e.g., location, condition, etc.).
- Anchor the 150 ft to 200 ft of $^{7}/_{16}$ in. (11 mm) kernmantle main wide-area search rope outside the entry point.
- Assign search positions. The assignment of RIT search positions is based on training procedures, accountability, safety, and keeping the operation as simple as possible. The following positions are recommended:

A. *Position #1* (RIT officer). Position #1 would assume the responsibility of the search team leader and should have possession of the thermal imaging camera. Position #1 will have the flexibility to move along the main wide-area search rope as needed to assess the interior layout, structural conditions, heat levels, and maintain a face-to-face accountability with the RIT search team if needed. It is recommended that Position #1 move into the search area with the other search positions to help direct the search, assess the level of risk, monitor SCBA air consumption, maintain communications with the anchor position, and assist the search operation with the thermal imaging camera (if available). It must be noted that there may be extreme situations (e.g., high risk, heavy smoke, etc.) at Position #1 when the RIT officer will carry, deploy, and control the search rope.

B. *Position #2* (Rope bag). Position #2 will deploy using the 150 ft to 200 ft of main search rope and also have each member carry a secondary rope with a maximum length of 50 ft. This secondary rope should have a clip or carabiner on the running end that can be attached quickly to the main search rope. Position #2 will search aisles, open areas, rooms, and any other target areas where a firefighter victim maybe located. Position #2 will initially team up with the RIT officer, but as additional search positions are set up, other firefighters assigned to the search position will work in teams of two, as ordered by the RIT officer.

C. *Position #3* (SCBA/Rope control). Position #3 should be a RIT member situated on the main search rope at a point where search team members are using the remaining main search rope or a secondary search rope to search aisles, open areas, stairwells, machinery rooms, and offices. Position #3 must have a large, hand-held flashlight pointing in the direction of the search team members, thereby attempting to maintain a visual contact with their movements by following their reflective turnout gear striping and hand-held flashlights. If possible, Position #3 should also have a thermal imaging camera to monitor the search and fire conditions in the search area. Because Position #3 will also be assigned the RIT SCBA air system, it will be difficult to move swiftly to search. It would be advantageous for Position #3 to hold a more stationary position on the search rope and remain readily available for the RIT SCBA air if needed.

D. *Position #4* (Tools/Search). Position #4 can carry additional rope and tools and assist with the search operations while being able to position an additional member at the search team entry point. If staffing limits the RIT to four members, then Position #4 can either remain at the search area entry point (as Position #5) or with search operation depending upon the RIT officer's decision.

E. *Position #5* (Entry). Position #5 is responsible for securing the main, wide-area search rope at the entry point, team accountability, setting up the entry point search light, assessing fire and structural conditions, and maintaining communications with the interior search team.

4. Check RIT equipment. The RIT officer must verify that all of the necessary equipment is accounted for and in working condition. What makes this search operation so different from the RIT single- and multiple-entry point searches is the depth into the building to which the search will extend. If a RIT member does not have a "working," hand-held flashlight, a radio that is on the right channel, or a secondary search rope, it is too late in many cases to crawl back some 100 feet in heavy smoke conditions to retrieve another piece of equipment. It endangers the RIT member, the other team members, and the lost or trapped firefighter they were sent to rescue.

5. The RIT officer must check the SCBA air levels of all RIT personnel prior to entry. It must be reminded that this is a team operation requiring great discipline. As soon as one SCBA low-air alarm activates during the search, the search operation in most cases must be discontinued until the SCBA air is refilled, and then the search can be reorganized and continued. If conditions warrant the use of a search rope, then all RIT personnel must have adequate SCBA air at all times.

6. Size up the interior layout of the potential search area from the entry point. The use of a thermal imaging camera, if available, is recommended for this task. The RIT officer should attempt to detect any hazards or difficulties such as truck wells, pits, holes, areas of collapse, entanglement problems, confusing aisles of high-rack storage, machinery, and overall fire conditions before committing to the search.
7. Set up (if possible) a 500-watt generator light shining into the entry point. If a light this size is not possible, a large, hand-held flashlight is recommended as opposed to a smaller, personal hand-held flashlight. Setting up a light at the entry point is done not only to illuminate the entry point but also to serve as a visual landmark for exiting.

RIT Wide-area search patterns. Wide-area search patterns will be dependent upon three main factors:

- Information concerning the location of the victim(s)
- The overall building layout: interior walls, number of rooms, arrangement of furniture, aisles, merchandise, rack storage, heavy machinery, and general housekeeping
- The amount of debris from fallen materials (e.g., ceiling tiles and grid work, high-rack paper bails, etc.)

Any one of these factors can determine and change the type of search pattern that is to be used. The key is to be trained and versatile so that all risk is calculated, but the search will still be able to overcome many of the obstacles in an effort to reach the firefighter victim(s). It must be reiterated that any rapid intervention search pattern must be basic, rapidly deployed, understood by all rescuers, effective, and as safe as possible (given the conditions).

Search patterns. If a Mayday is called by a lost firefighter in a 300,000 square foot building and the victim's location can somewhat be determined (via PASS alarm, yells, or hand-held flashlight), the fire conditions and the stability of the building will determine not only the type of search but how far the search can commit into the building. The following are the various search patterns used in a wide-area search. Please note the similarities as well as the differing aspects of each.

- *Wall search pattern.* The wall search must be conducted along a wall, allowing the search to extend from the wall using the main or secondary search rope(s). In determining the need for a wide-area wall search versus a different type of search, the fire and structural conditions and the amount of victim information will be the determining factors. A wall search can be fully deployed into the building if the fire conditions are heavy or if they consist of cold smoke. Once entering the building, an exterior wall will serve as the starting point to allow the rapid intervention officer to perform a reconnaissance and gather information for additional search teams. The walls will generally provide a better idea of the layout of the search area and another reference guide to the way out of the building for the rescuers. Many firefighters have been found on or near an exterior wall unable to find a window or door for escape. Generally, fire service training practices have encouraged firefighters who become lost and disoriented to attempt to locate an exterior wall with the hope of finding a window or door for escape. So with good reason, when there is little information as to the location of the victim(s), the walls are good starting points for a search operation.

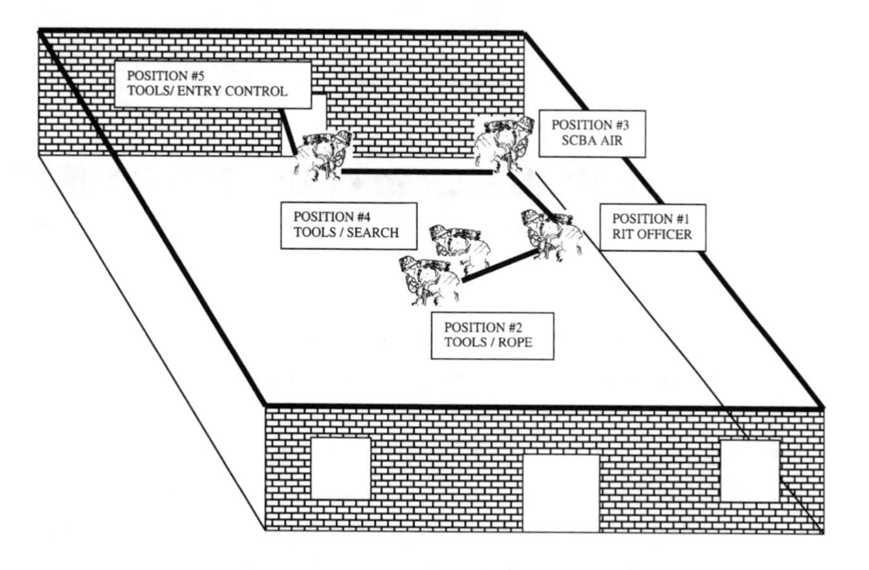

Fig. 6–12 Rapid Intervention Wide-Area Search

- *Sweep search pattern.* Once the RIT has reached the sector to be searched, a sweep search can be conducted by Positions #1 and #2 working from Position #3's staged position. It must be reminded that Position #3 is responsible for the RIT SCBA air system and tools, which inherently cause some difficulty in rapid search moves. In addition, this staged position is also the RIT officers "checkpoint" for emergency SCBA air if needed. As the RIT extends into a sweep search, the "sweep" will allow them to search around walls, furniture, machinery, and many other obstacles as needed.

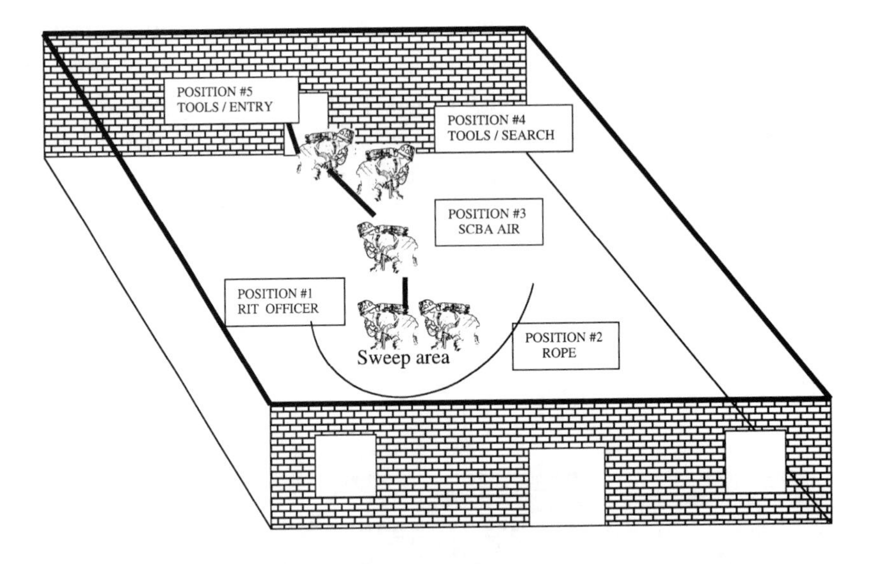

Fig. 6–13 Sweep Search Pattern

- *In-line sweep search.* The in-line sweep search is very methodical sweep search of a wide area for a missing firefighter. As searchers connect their secondary search ropes to the main search rope and deploy outward, they space apart approximately a distance of an arm's length. Once in position, the searchers search in-line on the secondary search ropes as needed to cover the search area as thoroughly and quickly as possible. Once the searchers have returned to the main search rope (in a coordinated manner), the RIT officer then moves the search team forward or can deploy them in the opposite direction.

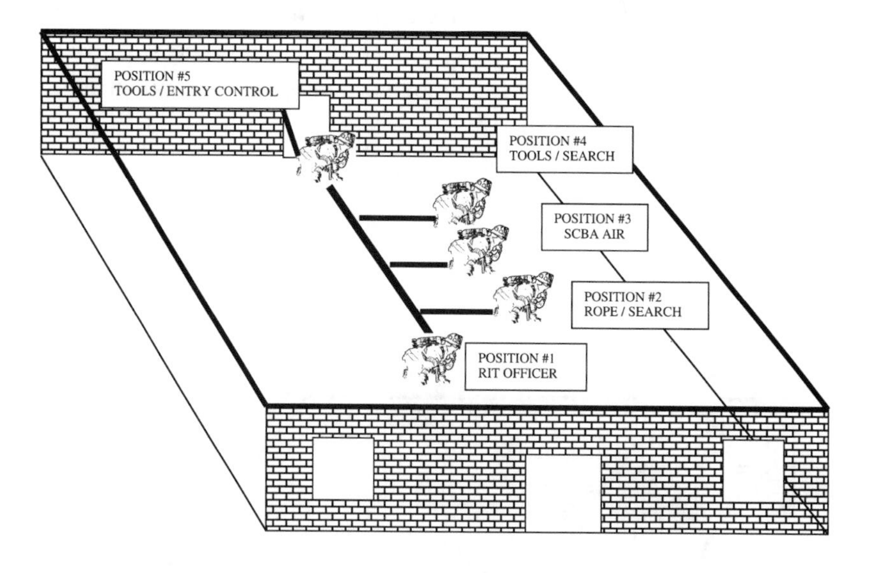

Fig. 6–14 In-Line Sweep Search Operation

- *Shoot-out search pattern.* The shoot-out search pattern can also be used by either the main or secondary search ropes. This search pattern is designed for straight-path searches along walls, aisles, hallways, stairwells, between machinery and racks, and in confined spaces due to collapse. It is one of the simplest search patterns and, depending on the size of the space to be searched, allows one or more rescuers to proceed forward while deploying a search rope. Once the search has been completed, the rescuers pull the rope taut and return to their starting point, move to another position, and deploy into another aisle or hallway. If there is a need to search a channeled and/or confined area, the shoot-out search pattern is one of the most feasible searches to allow a rescuer(s) to search quickly. It is important to realize that searchers can advance too far very quickly into dangerous and unstable conditions. It will be important for the RIT officer to oversee the shoot-out search patterns at all times. Once the victim is located, the determination must be made if a rescue drag can be performed or if the victim requires extrication. If the victim can be dragged, it will be important for Position #3 on the main search rope to pull and keep the rope taut at all times as the rescuer returns to the main search rope. If the firefighter victim must be extricated, it will be important for the rescuer to tie off the search rope as near to the rescue area as possible, again pulling the search rope as taut as possible. In the event additional rescuers will be needed or conditions deteriorate, the rescuers can firmly grasp the search rope and evacuate the building without the rope becoming limp and no longer able to provide guidance to the exit.

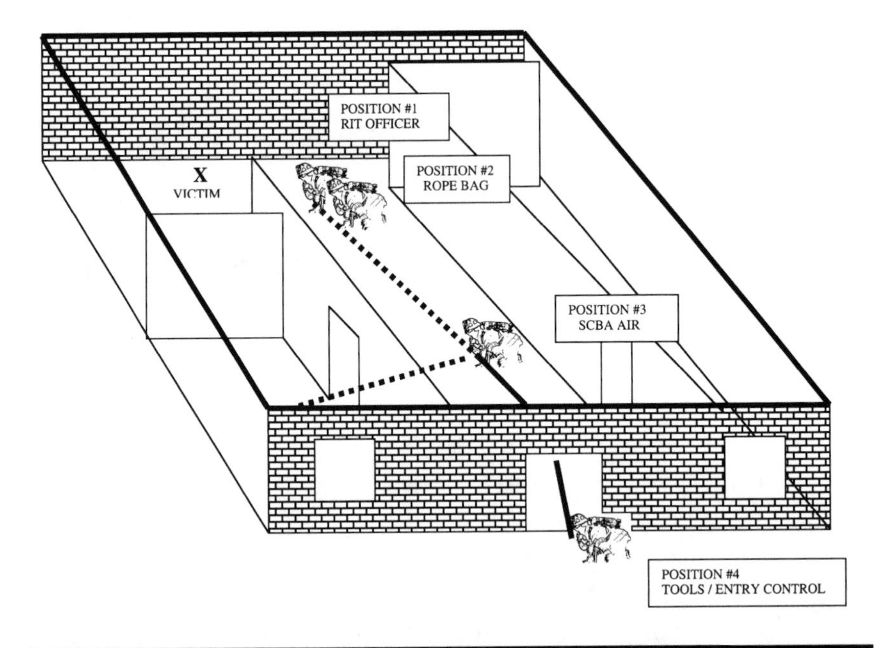

Fig. 6–15 Shoot-Out Search Pattern

RIT Wide-Area Search Operation Checklist

- Determine RIT wide-area search entry point(s).
- Divide and sector building.
- Determine need for a hoseline to accompany search operation.
- Assign RIT wide-area search positions.
- Obtain the most recent victim information.
- Anchor main search rope.
- Check basic equipment:

  - One 150 ft of 11 mm or $^{7}/_{16}$ in. kernmantle of main search rope
  - Two 35 ft to 50 ft kernmantle secondary search ropes
  - Four (minimum) portable radios on the same dedicated channel
  - Spare SCBA or alternate air supply system
  - Hand tools
  - Thermal imaging camera (if available)
  - RIT personnel check of SCBA air cylinder pressure
  - RIT personnel all have hand-held flashlights

- Size up interior layout (e.g., fire conditions, high-rack storage, pits, machinery, etc.).
- Set up search light at entry point (500-watt generator light if possible).
- Brief RIT on search plans, and confirm team accountability.
- Document entry time with Position #5.
- Deploy RIT wide-area search team.

7

Firefighter Rescue Removal Methods

Firefighter rescue removals are skills that should be learned at the firefighter entry level of training. These skills are as important as firefighter self-survival methods, in that a firefighter removal might have to be executed by any firefighter at the scene of a fire no matter what their rank or responsibility. There have been many cases when chief officers and pump operators alike have had to raise ground ladders and position aerial ladders to windows and fire escapes to rescue escaping firefighters. Rescue removals can include floor drags, haul systems with rope, or ladder removals. It is important at this time to review reasons why firefighters become victims at fire incidents:

- SCBA malfunction or error in use
- Firefighter becomes separated and lost, thereby using any remaining SCBA air
- Firefighter is trapped, pinned, or entangled due a structural collapse
- Firefighter becomes separated and lost as a result of disorientation
- Firefighter becomes injured, entangled, and/or trapped
- Firefighter suffers a medical emergency such as a heart attack, stroke, and/or heat stress

Note that the two most important considerations when determining how to rescue and remove a firefighter victim are:

- **The location of the firefighter victim.**
- **The type of fire and current structural conditions.** The firefighter rescue removal method(s) to be used will strongly depend on where the victim is located in the building and how the victim got there. For example, understanding how a firefighter fell into a basement and whether or not there are stairs available or in strong enough condition to be used to attempt a removal will determine which method(s) can be used. This is a time when firefighters must think "outside of the box" not only in terms of self-survival as discussed in chapter 4 but also in the context of firefighter rescue. Any firefighter rescue removal method is subject to change, modification, or additions. This is a time when experience, ingenuity, and teamwork will be required to develop a Plan B, or even a Plan C, to back a possibly unsuccessful Plan A.

Experience has shown during so many documented firefighter fatality incidents that Murphy's Law, which simply states that anything that could go wrong...will go wrong, has happened repeatedly. Radio failure, burst hoselines, structural collapse, loss of accountability, and so many more problems have been known to occur when a firefighter was reported to be missing, lost, trapped, or when a victim called a Mayday. Using the earlier example of the firefighter trapped in the basement, if Plan A was to pull the victim up the stairs and it was found that the stairs were too weak to be used for the rescue, what is Plan B? Plan B can involve shoring the staircase from underneath, and a subsequent Plan C can involve using a ground ladder to bridge over the top of the stairs to distribute the weight. There may even be a possibility that both Plans B and C can be combined for extra stability to handle a live load of more than 600 pounds of the rescuers and victim combined. A rescue basket can be added with a simple haul system to slide it up the ladder.

But never forget the ever-present Murphy's Law. What if the stairs still fail or if there is not enough room to negotiate the victim up the stairs? Yes, you guessed it, what is Plan D? At this point, it is clear that the odds are against a successful rescue and removal of a firefighter victim due to the extended time that has elapsed. Added to this are the worsening fire conditions, collapse potential, SCBA air consumption, condition of the victim(s), and the fatigue and emotional strain of the rescuers.

A Case Study in Frustration

In Chicago on November 22, 1976, a five-alarm fire occurred at a Commonwealth Edison power plant where a quarter-mile long section of a burning coal conveyor collapsed onto and through the 85 ft roof. At the time of collapse, Firefighter Walter Watroba from Engine Company #13 was working a hoseline with two other firefighters. Although all three were initially trapped, two were able to escape and were immediately rescued, but Firefighter Watroba had become pinned by tons of twisted steel and concrete on the roof of the eight-story power plant. Immediately after the tremendous collapse of the conveyor system, it was initially thought that everyone was accounted for as companies repositioned master streams outside of the debris area, but then a firefighter's hand-held flashlight could be seen shining through the smoke on the edge of the power plant roof. When Firefighter Watroba was found, only his head could be seen as the rest of his body was covered in debris. He was in a seated position pinned against the inside of an exterior wall. As the large 10 ft X 10 ft square conveyor system sheared completely, it fell across his lap crushing both legs. Firefighter Watroba had immediately suffered internal abdominal injuries, his left leg was somewhat pinned and was unsalvageable due to the extensive injuries, and his right leg was severely pinned under the tons of steel and concrete.

The initial rescue effort to reach Firefighter Watroba involved the use of a 55-ft snorkel, fully extended from Snorkel Squad #1, and a 30-ft straight ladder that was placed in the basket to reach the 85 ft high roofline. As the ladder was secured in the snorkel basket, the first rescuer to reach Watroba found that the ladder tip was several feet short of the roofline making the reach for the victim a dangerous balancing act. It was also found that a hoseline was needed immediately, because the fire in the collapsed conveyor chute was moving up rapidly toward Watroba. This required the rescuer to descend back into the snorkel basket and take a 1½-in. hand line back up to the roof. Once at the roofline, he attempted to flow water into the coal chute to beat the fire back from the victim, but then he started to lose the nozzle due to the unsure footing on the ladder and ice. The rescuer then handed the hoseline to Watroba, and he directed the stream into the chute finally knocking down much of the fire. Once some of the debris, smoke, and steam cleared, it was found that Watroba was trapped on an 18-in. wide ledge of what was left of the roof. Rescuing firefighters immediately used part of the wall to secure Watroba with rope so he would not fall into a wide crevice at his immediate left side that dropped off nine stories.

During the initial rescue effort, no more than two rescuers were able to position around Watroba. As the subsequent rescue of Firefighter Watroba started, the following complications plagued the rescuers for more than $7\frac{1}{2}$ hours:

- Time. Constraints in time due to the victim's medical condition were critical as he was suffering from shock, crushing injuries, smoke inhalation, exposure to 28°F temperatures, high winds, and snow.
- Secondary collapse. The structure was weakened after the first major collapse, and, therefore, whenever rescuers attempted to use any part of the building as a base for a hydraulic or pneumatic tool, there was the risk of further collapse.
- Time of day. The collapse and rescue efforts occurred during the night hours in a location where lighting was difficult to supply.
- Weather. The rescuers had to operate in constant cold, snow, and ice over extended periods of time.
- Transporting the medical and rescue equipment. Hydraulic, pneumatic, and various hand tools had to be moved up from the ground to the rescue area, which was 85 ft high.
- Availability of equipment and tools. There was very little ability to stage any tools in the rescue area, which caused great delays in the rescue plans because the tools had to be transported up and down via ladder.
- Communication. Although communication could be accomplished via radio, it was often incomplete and restricted. This caused extreme confusion and frustration among the chief officers who could not reach the rescue area and the rescuers who had difficulty in trying to explain any progress or set-backs throughout the rescue effort.

After the loose debris was removed, the victim was asked if he could possibly free himself, but it was not possible. A Hurst tool with spreading tips along with 20 and 30-ton jacks were moved up to the rescue area. The rescuers wanted to lift the massive coal chute up and off Watroba's legs. However, it was found that between the weight of the chute and instability of the building's exterior wall, removing the chute would increase the possibility of a greater structural collapse. Once the Hurst tool was sent back down, high pressure air bags were then sent up. Again, with the same complication of not having a solid structural base to work from, the air bags did not work. The next attempt involved the use of an air chisel. The attempt to cut out the section of the coal chute around Watroba's legs failed because of the density of the concrete that surrounded the steel lined chute. As it was chiseled, the concrete failed to break off in chunks; instead, it would only turn into a fine powder. During much of this time, the victim was given morphine for pain.

Approximately five hours into the incident, desperate ideas started to emerge. The use of cranes to lift the chute was considered but discounted because of the difficulty and time of getting them into position. Then the use of a helicopter lifting a sky crane was considered, but it was also discounted because of the size and weight of the collapsed coal chute. Along with the desperate ideas came frustration, fatigue, and emotional reactions among the rescuers because of the constant failures in rescuing Firefighter Watroba.[13]

Almost seven hours into the incident, it was decided to attempt to rescue the victim by amputation. A fire department doctor, although not used to the hostile and dangerous conditions, quickly committed to the operation. The amputation above the right knee took only a few minutes, but by that time the shock, exposure, and trauma proved to be too much for Firefighter Watroba. Sadly, he did not survive the ordeal and succumbed to his injuries before reaching the hospital approximately $7\frac{1}{2}$ hours after becoming trapped.[14]

This very sad incident is meant to relate to all rescuers that in spite of the most advanced rescue tools, experience, and training, there will always be situations where nothing will work to successfully save firefighter victims.

Fig. 7–1 The Day After the Fire, a View of the Com Ed Plant and Part of the Collapsed Coal Chute (Ken Wood)

Fig. 7–2 Chicago Firefighter Walter Watroba Being Removed from the 85-ft High Rooftop of the Burning Power Plant (Chicago Fire Department)

Initial Firefighter Rescue (IFR) Steps

The type of fire conditions will strongly determine the type and timing of the rescue removal method(s) to be used. Any rescue removal of a firefighter victim should be as fast as humanly possible, but when fire conditions are so extreme and a removal attempt is made, there may not be any particular learned method that can be used. At such times, just a "grab 'n go—whatever works" method will be necessary. Imagine crawling up a set of stairs in a house and hearing a PASS alarm near the top at the landing. While approaching the top step, fire is lighting up over your head, and you see the firefighter victim within 3 ft of your reach. Whether a hoseline is available or not, the conditions can turn fatal at any second. It is at this point that the rescuing firefighter will grab any part of the victim, pull toward the stairs, and slide or tumble down the stairs to outrun the dropping heat. Escape is made without using any particular removal method that the rescuer had trained for. In such circumstances, the important thing is to rescue and survive in whatever way possible.

The following terms can be used to explain levels of fire conditions:

LIGHT	Little or no heat and light lazy smoke.
MODERATE	Warm heat with smoke almost to the floor and becoming thicker.
HEAVY	Hot (a "biting" heat to touch) with a heavy push of thick smoke.

With heavy fire conditions

1. **"Grab and go."** Get yourself and the victim out of the life-threatening conditions immediately with a hasty grab
2. **Rescue drag the firefighter victim.** Try to rescue drag yourself and the victim to a safe area (e.g., floor below the fire, nearby uninvolved apartment, or protected stairwell), or hand off the victim to additional rescuer.
3. **Call for help.** Verbally call or transmit a Mayday distress call via radio once out of extreme life-threatening conditions. The Mayday transmission should cause the following actions to occur:

- Shut down all fireground radio traffic
- Activate the RIT, RIT sector officer, and RIT support teams
- Initiate an accountability role call of all fireground personnel

Depending upon the complexity of the Mayday incident:

- Request an additional alarm
- Assign a RIT Rescue Sector

With light or moderate fire conditions

1. **Turn the firefighter victim faceup.** By way of an informal count, many NIOSH fatality investigations have revealed that a majority of firefighter victims have been found lying facedown. It could be assumed that such victims were staying low from the fire conditions as they were crawling, attempting to see where they going in an effort to escape, protecting themselves from the heat, or from falling debris. If the firefighter victim is found facedown, it will be extremely important for the rescuer to turn the victim onto their back (faceup).

   This one isolated move can account for the victim's increased chances of survival for the following reasons:

   - *Reach to silence the PASS alarm.* Since "stand-alone" PASS alarms can be mounted on the SCBA waist harness, which can allow them to slide on the harness and make them difficult to reach, there is a better chance of reaching the PASS alarm by turning the victim faceup. Integrated PASS alarms are generally positioned on the SCBA shoulder harness making its accessibility much easier when the firefighter victim is turned faceup.

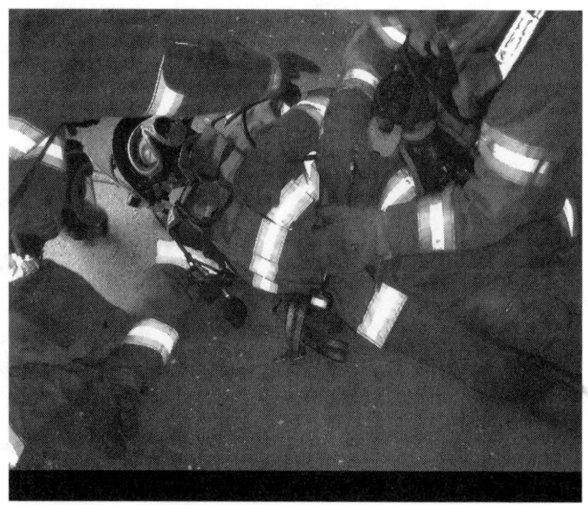

Fig. 7–3 A Firefighter Victim in a Faceup Position After Being Turned by a Rescuer (Kolomay)

   - *Reach the remote radio microphone.* There will be times when the firefighter victim has a radio and the rescuer does not. If the victim has a remote that is positioned near the collar, it can be used by the rescuer.
   - *Identify the victim.* If the victim is not identifiable by a tag or printed name on the turnout coat, it can be possible to identify him or her faceup by helmet markings or facial recognition.

- *Ascertain if the SCBA facepiece is still in place.* It will be possible to see if the victim is still wearing the SCBA facepiece or not.
- *Prevent drowning.* Many firefighters who have been found unconscious have been threatened with drowning from accumulated water from hose streams and sprinkler systems. Turning the victim faceup will lift their face from any water that may be present.
- *Adjust and reach the SCBA harnesses.* Whether you plan to drag the victim by means of the SCBA harness or you simply want to remove the SCBA harness, if the victim is turned faceup, then the buckles that adjust the tightening and loosening of the SCBA harness, are now visible and easier to reach. When the victim is positioned faceup while lying on the SCBA cylinder, it will cause the victim's body to be tilted with a high side and low side. Reaching for an SCBA harness, arm, or leg from the "high side" will make the rescue faster and easier.
- Drag easier. Due to the fact that the firefighter victim has been turned faceup, the SCBA air cylinder will be set on the floor. When the rescuer(s) drag the victim, the air cylinder will tend to act like a runner on a sled allowing the victim to be dragged much easier and clear through such debris as plaster, lath, wallboard, etc.

2. **Reset the PASS alarm.** The purpose of resetting the PASS alarm is so the rescuer, who is radio equipped can transmit via radio without having to compete with the approximately 100-decibel alarm sound. With the PASS still activated, it is probable that the radio transmission for help and direction would have to be repeated, costing time and adding frustration. In addition, if the transmission does make it through to the IC and the RIT, the noise from the PASS could distort the message, resulting in a misinterpretation of the information. This could possibly send the RIT to the wrong entry point and cause many other such problems. If for some reason the PASS

Fig. 7–4 The Firefighter Tilted on the SCBA Cylinder Giving Way to a High and a Low Side for Easier Reach of the Shoulder Harness, Arm, or Leg for Rescue (Hervas)

alarm reactivates during the rescue and it cannot be reset, at least the initial transmission for help was successful, and help is on its way.

Stand-alone PASS alarms can be turned off completely. However, in the case of a PASS alarm that is integrated into the SCBA, the PASS alarm cannot be turned off without turning off the victim's SCBA air supply from the air cylinder. If the victim is not moved, the integrated PASS alarm will go into pre-alert repeatedly and must be reset. This is most important if the firefighter victim is trapped, requiring extrication, and the SCBA is activated supplying air. It will be important to keep the PASS reset for clear radio communications and to maintain a lower level of anxiety.

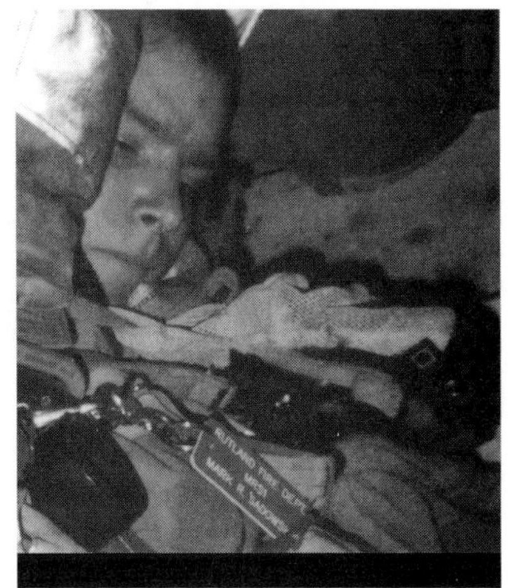

Fig. 7-5 Identification by Facial Recognition, Helmet Markings, or Identification Tag (Hervas)

3. **Transmit a call for help.** Transmit via radio "Mayday! Mayday! Mayday!" Repeated notification of a Mayday distress call will leave little doubt whether anyone is in trouble, whereas one Mayday distress call may not be heard through the raucous radio transmissions after a sudden collapse.

4. **Confirm the Mayday distress call.** As a firefighter rescuer, it is important that once the firefighter victim is located, the Mayday situation is confirmed by relaying:

   - *Victim's location.* Indicate if the victim is, for example, in a basement at the rear stairs, on the second floor–Sector A, or in Stairwell B on the 14th floor.
   - *Victim's condition.* Determining whether the victim is conscious, unconscious, entangled, or pinned will provide the incident commander with an initial idea of how difficult the rescue will be and how much help will be needed.
   - *Victim's identification.* If possible, indicate the victim's name, assigned company, or the fire department the victim is from. It is important for all firefighters to clearly place their name on all of their personal protective equipment, not only for equipment that is "lost and found" but also for firefighters that are "lost and found." Some fire departments have also specified their turnout coats to have the firefighter's name printed on the back of the coat or to affix small brass identification tags on the coat hook or zipper. The identification of the victim is very important for accountability reasons.

5. **Check the SCBA facepiece.** If the victim is not wearing a facepiece, it will be a judgment call as to whether or not a facepiece should be replaced based on the fire conditions, the victim's physical condition, and time. If the victim is still wearing the SCBA facepiece, it will be necessary for the rescuer to check to see if it is still supplying air. If this is not done, there is a good chance that the victim will suffocate in the facepiece. To check for airflow, pull the bottom of the facepiece away from the victim's chin to break the seal, or activate the bypass or purge valve and then attempt to listen for a positive pressure "blow" of air from the facepiece. If you do not hear the "blow" of air, remove the mask-mounted regulator from the facepiece or the breathing tube from the regulator depending on the type of SCBA being used. Although the room air might be warm and smoky, it will at least provide for some air exchange until the victim is rescued into clean air or given new SCBA air.

Fig. 7-6 Checking for Airflow in the SCBA Facepiece (Hervas)

Although it is not wrong to remove the entire SCBA facepiece, it is not recommended. If trained properly, removing the mask-mounted regulator or breathing tube is actually faster than fighting to remove the whole facepiece, which is covered by a protective hood, a helmet with a chin strap, and sometimes a facepiece neck strap. Leaving the facepiece on can offer the firefighter victim some additional facial protection from debris and heat, possibly provide some respiratory protection, and remain available for shared air operations in the event the victim is trapped and requires a difficult rescue. If the facepiece remains on the victim's face, the victim can easily be given SCBA air.

Initial Firefighter Rescue Steps with Light to Moderate Fire Conditions

Locate and turn the firefighter victim(s) faceup.

↓

Reset the PASS alarm (if activated).

↓

Transmit a Mayday or confirm a Mayday.

↓

Check the SCBA facepiece.

↓

Remove the victim.

Each initial firefighter rescue step is designed to increase the chances of survival for the victim. The importance of each step can be looked at from a perspective of how the victim's chances of survival might diminish by not performing each step as needed:

- If the rescuers do not call for help, it will be that much more difficult to perform a rescue and the risk becomes increased.
- If the firefighter victim is not turned faceup, it will be very difficult to perform any of the rescue steps.
- If the rescuers do not reset the PASS alarm, there is a good chance communications will be obscured. Transmissions will be scrambled, unclear, leading to poor directions.
- If the SCBA was not checked for airflow, there is a good chance the victim will suffocate. It would be sad to perform a successful high-risk rescue only to find that victim died from suffocation.

With adequate training, all of the initial firefighter rescue steps performed prior to removing the firefighter victim can take less than 10 seconds. An acronym that can assist in the recall of the initial firefighter rescue steps is **FRAME**.

Find the firefighter victim.

Roll the victim over onto the SCBA cylinder (if the victim is facedown).

Alarm, reset the PASS alarm.

Mayday distress call.

Exhalation of SCBA air; check the SCBA facepiece for exhalation of air.

Communication among Firefighter Rescuers

Communication among firefighters on the fireground during normal operations in fighting a fire is difficult and even more so during a Mayday incident. During this situation, communications are extremely difficult if not impossible at times. It will be most important for any rescuer and RIT member to be trained to coordinate the various rescue skills with strict and instinctive communication skills. If two or more rescuers do not coordinate their rescue efforts, their odds of a successful rescue will decrease tremendously. By way of an example, what has experience shown when an unconscious firefighter must be lifted vertically to a 36-in. high windowsill? If the two rescuers do not simultaneously conduct the lift, one of the two rescuers will end up expending whatever remaining energy they had trying to lift the victim. The outcome is that the victim will not even leave the floor and an immediate second lift "to try it again" will be out of the question.

"Ready? Ready. Go!" rescue command

To coordinate any of the firefighter rescue skills with multiple rescuers, the command of "Ready? Ready. Go!" is given. Either one of the rescuers can initiate the first "Ready?" Once the first rescuer states "Ready?" the second rescuer would respond "Ready." Then the first rescuer would give the "Go!" command to simultaneously conduct the lift, carry, spin, or drag with the second rescuer.

Firefighter Rescue Drag Methods

There are a variety of rescue methods that can be used to remove a victim from the fire scene. They include drags: drag and carry, side-by-side drag, push/pull drags, as well as other types of drags for various situations. All of the firefighter rescue drags introduced in this book are designed for last ditch, life-saving situations, wherein the firefighter victims who are located are in a state of unconsciousness and in immediate need of rescue. Each of the rescue drag methods are designed for any firefighter to use regardless of their position or assigned duty at an emergency incident. A major consideration that has been reemphasized in each chapter has been the firefighter victim's weight and how to maneuver it across floors, up and down stairs, through holes and confined spaces, off of roofs, and out of entanglement situations. Each of the firefighter rescue drags have been designed for rescuers who are physically fatigued, emotionally stressed, limited in help, low on SCBA air, and having to contend with fire, smoke, and debris conditions. There is no question that the difficulty of the drag will be determined by the condition of the rescuer and the victim as well as the level of risk involved. Given light to moderate fire conditions for each of the firefighter rescue drags, the initial firefighter rescue (IFR) steps must be executed to perform the drags in the most effective manner and increase the chance of survival for the victim being rescued.

One firefighter rescue drag method with SCBA

This rescue drag method is to be performed by one firefighter rescuer in the event:

- A firefighter's partner becomes an unconscious victim and must be rescued
- Two firefighters locate a firefighter victim. One firefighter must clear out a window or doorway for escape as the other firefighter drags the victim to that window or door
- Firefighters locate and remove two or more firefighter victims, allowing for only one rescuer for each victim

As the rescuer stays as low to the floor as possible, the rescuer performs the IFR steps, and the rescuer then grasps the "high side" SCBA shoulder harness. When attempting to grasp an SCBA shoulder harness at any time, the rescuer can best grasp the harness by reaching under the shoulder harness at the top of the SCBA back frame first. Generally, at the top of any SCBA frame, there is a gap between the victim's shoulder and the harness; this gap will allow the hand to grasp the harness easier.

As a special note, depending on the type, age, and condition of the SCBA, it might be necessary for the firefighter rescuer to take the excess SCBA shoulder harness strap and double it over so it will not extend when the SCBA is pulled. If the SCBA shoulder harness unintentionally extends, there is a chance the firefighter victim could be pulled out of the harness.

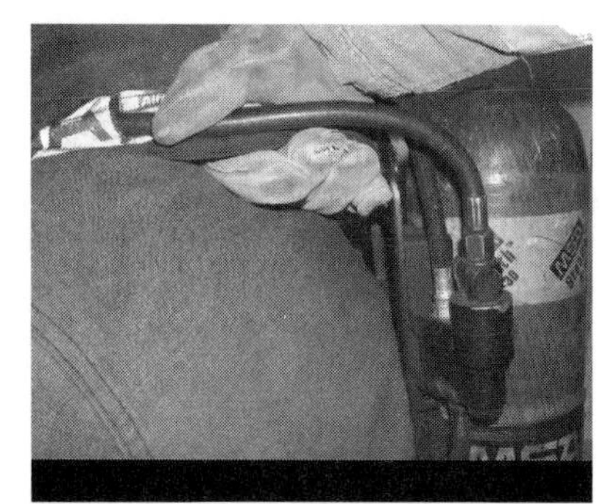

Fig. 7-7 Grasp the SCBA Shoulder Harness at the Top of the SCBA Back Frame Between the Shoulder and the SCBA (Hervas)

It may not always be necessary for the firefighter rescuer(s) to stay low to the floor, especially if the fire conditions are light and the building is, structurally, in good condition. For that matter, grasping the SCBA shoulder harness and pulling the victim from an upright position is a much easier and more forceful pull than from a crawl position.

Fig. 7-8 Firefighter Rescuer Remains Low to the Floor Performing a One–Firefighter Rescue Drag (Hervas)

Because the crawl position is more difficult, it will be necessary for the firefighter rescuer to use counterweight and leg-power to pull the victim. As one hand grasps the SCBA shoulder harness, the other will sound the floor for holes, stairs, and debris. It will be important for the rescuer to kneel with the outside leg for balance, and use the other leg to do the heavy work by pushing off the floor so the victim could be pulled. While the victim is being pulled, the firefighter rescuer can also use upper body weight to lean in the direction of the pull to help counter some of the victim's weight.

Firefighter rescue drag and carry methods without SCBA

Several years ago when firefighter rescue and survival training became very prevalent within the national fire service, many of the rescue drags and lifts concentrated on the use of the SCBA harness as a "rescue harness" to assist in the firefighter victim removal. The belief at the time was that there were very few instances, if any, when a firefighter becomes a victim in a fire building when they would not have an SCBA on; this assumption was proven wrong.

On March 14, 2001, at 4:54 P.M. in Phoenix, Arizona, the Phoenix Fire Department responded to a reported fire behind a strip shopping center consisting of a number of small stores, a large hardware store, and a grocery store. The fire was located in the rear storage room of the supermarket with heavy smoke and moderate heat conditions.

Phoenix Firefighter Bret Tarver, who had assisted in advancing a hoseline into the rear of the supermarket, had become low on SCBA air. After informing his officer, the company began to exit several firefighters, including Tarver. However, Tarver became disoriented after hitting a wall and then became separated and lost. Firefighter Tarver was found approximately 60 ft from the nearest door, somewhat entangled, half-way under a table, and unconscious with worsening fire conditions. The RITs then confronted new challenges when the firefighter victim had to be dragged through and over fallen debris and around aisles without the use of the SCBA as a rescue harness. The Phoenix firefighters involved in the rescue were all extensively trained in firefighter rescue and rapid intervention operations, but as mentioned earlier, training on firefighter rescue drags without an SCBA harness had not evolved at that time.

Tarver's SCBA having been removed, the first RIT was at a great disadvantage and experienced difficulty in dragging him. After moving the victim about 15 ft, the second team performed a rescue drag but had to utilize the victim's turnout clothing to pull him. The grab points on the turnout coat did not allow for a sure grip as the turnout coat would slide and would become snagged and start to turn inside out. As the third RIT moved into position to continue the attempted rescue of Firefighter Tarver, they found few grab points to pull from and resorted to the uniform tee shirt. Firefighter Tarver was larger in stature and weighed more than the average firefighter, and the tee shirt tore away. The rescuers then tried to grasp his arms and wrists, but this only made the rescue more difficult.

Compounding the rescue drag, there were constant entanglement problems from the aisle shelving units, wooden pallets with protruding nails on the aisle floor, and fallen ceiling debris. The rescuers were not only required to stop and disentangle the victim but also to lift the victim up and over the debris. As the fourth and final RIT took over, they, like the previous teams, had to contend with not only slipping on the tile floor that was wet from hose streams but was also coated from cooking oil products that had fallen from shelving. They had to contend with all of this while trying to move the victim through confined spaces.

In all, four RITs were needed to rescue Firefighter Tarver over an estimated time of 31 minutes. As stated in the final report conducted by the Phoenix Fire Department and IAFF Local 493, "During the rescue, the movement of Firefighter Tarver was made extremely difficult by a multitude of hindrances that included limited visibility, water from the firefighters' hoses, heat conditions, warehouse obstructions, and falling debris. In addition, rescue efforts were severely compromised by the limited remaining air supply in the rescuers' breathing apparatuses, debris caught in Firefighter Tarver's protective clothing and equipment, and Firefighter Tarver's physical size." This very disheartening experience and loss in Phoenix, Arizona demonstrated the importance of providing firefighter rescue drag methods and training for rescuing victims without an SCBA harness.[14, 15] It is also another example of how dynamic firefighter rescue and survival is, in that it will always be subject to change and improvement based on new equipment and experiences.

The following methods are options when SCBA is not available to act as a rescue harness:

Turnout gear drags and carries. Using the turnout gear is the least desirable method for a rescue, but it is an option if firefighters have been trained for this method of rescue removal. The key aspect of using turnout gear is knowing where and how to drag and lift the firefighter victim. The downside of turnout gear is the design of turnout gear itself. Closures that use hook and dee clasps, breakaway zippers, Velcro®, and the length of the coat itself can inherently cause the gear to fall apart during a drag or lift. Another adverse situation is how the firefighter victim is wearing the turnout gear. Firefighters come in various shapes and sizes and, in some cases, do not clasp or zip the closures or use suspenders when necessary. This results in the turnout gear virtually coming apart when pulled. Therefore, using turnout gear for rescue will require the rescuing firefighter(s) to use certain "grab points" as follows:

- The collar. Whether performing a one- or a two-firefighter rescue drag, using the collar will generally work well when dragging a victim. As long as the turnout coat can catch the victim under the arms, short drags might work. It will be important for the firefighter rescuer(s) to check that the turnout coat is fastened and not to let the victim's arms raise above the head, so the coat cannot either turn inside out and/or be pulled off the victim.
- The bottom of the turnout coat. When performing a two-firefighter rescue drag, where rescuers have to turn or lift the firefighter victim, it is best for both rescuers to grasp the outer shell of the bottom of the turnout coat (typically where the bottom reflective trim is located). It must be noted that if both firefighter rescuers grab a different part of the turnout coat when turning or lifting the victim, there is a good chance the maneuver will fail and the turnout coat will twist, turn inside out, or even be removed. For example, if one rescuer grabs the bottom of the turnout coat and the other rescuer grabs the back pocket of the turnout pants when having to lift the victim over an obstacle, chances are good that the lift might not be successful because the coat will twist around the victim and the rescuer will lose lifting ability as a result.

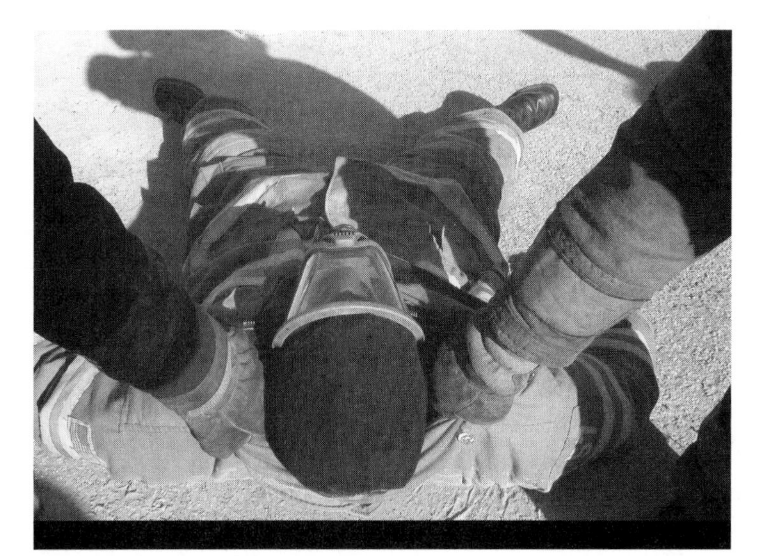

Fig. 7-9 Collar Grab and Drag (Hervas)

Turnout-gear carry. By having two firefighter rescuers use the collar and coat bottom grab points, while a third firefighter rescuer separates the victim's legs and grabs the knees using a "wheel-barrow carry," the firefighter victim can be lifted over obstacles and carried forward.

Class I waist-belt drag. There are waist belts available that are standard issue for many fire departments and can be used to secure a firefighter victim for rescue removal drags and lifting. If such belts are to be counted as a rescue harness, they should be tested and rated as a Class I harness for rescue. Although they can provide several secure "grab points" for rescuers, there are several disadvantages with using a Class I belt for rescue removals.

- Pulling the firefighter victim from the middle of the body is difficult. The victim is not balanced, will entangle, and will not negotiate around corners very well. Compared to pulling the victim from the shoulders or from around the chest, there is a much greater workload for the rescuer(s).
- When tying such a belt to the arms, legs, or ankles it does not always stay secure on the victim, especially when having to negotiate up or down stairs. In many cases the belt is too short to be adequately tied and secured around the victim's limbs.

Firefighter rescue with webbing harness

The use of the rescue webbing harness is the most effective method for removing a firefighter victim without an SCBA harness. In keeping with the intent of any firefighter rescue being a quick, "down and dirty" process, the use of Class III harnesses, haul systems, and the like are not realistic methods. There is, however, a rescue harness that has proven successful for firefighter as well as civilian rescues. This firefighter rescue harness is made of 1 in. tubular nylon webbing rated at 4000 lbs. tensile strength and is 20 in. long. The two ends of the webbing are tied with a ring bend knot (water knot), which is a specific knot for webbing material. A carabiner, rated at least 4000 lbs. tensile strength, is added to the webbing rescue loop, which will allow for many options in terms of securing, anchoring, harnessing, and hoisting. It is recommended that a non-locking "keylock gate" carabiner be used, since it has proven to be the easiest type to work with when working with limited vision and heavy gloves. Cities, including Chicago and Phoenix, have already been providing rescue harnesses with this design as standard issue to their firefighters.

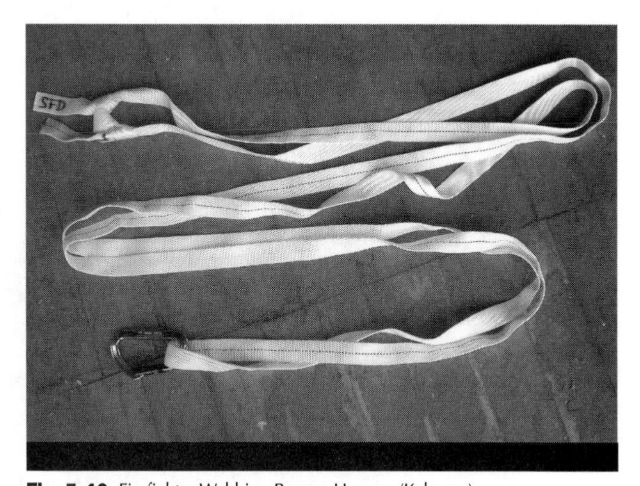

Fig. 7–10 Firefighter Webbing Rescue Harness (Kolomay)

Firefighter rescue webbing around the victim's chest. The steps to harness the victim around the chest must be done even under difficult smoke conditions:

1. Turn the victim faceup, if necessary.
2. Sit the victim up.
3. Remove the webbing from your turnout gear pocket, and grab the carabiner.
4. Feed the carabiner and webbing around the first arm, around the chest, and under the second arm to the victim's back.
5. Clip the carabiner onto the webbing at the victim's back.
6. Take up any slack, and place your hand into the harness.
7. Wrap your hand once or twice with the extra webbing to secure your grip.
8. Lay the victim back, and perform a rescue drag. Utilizing the rescue webbing, the firefighter victim can be removed using any of the methods that are used with an SCBA.

Fig. 7-11 Firefighter Rescue Harness Around Victim (Kolomay)

Fig. 7-12 Making the Webbing Hand Cuff Knot (Kolomay)

Firefighter rescue webbing handcuff knot. In the event the firefighter victim is entangled, caught in debris, or cannot sit up so the rescue webbing can be placed around the chest, an

Fig. 7-13 Firefighter Rescue–Webbing Handcuff Knot Being

alternate method of tying the webbing into a handcuff knot can be used. By placing the webbing onto the wrists or legs of the firefighter victim, the victim can be dragged a short distance to clear the debris, and the rescuers can resort to an easier side-by-side or push/pull method.

Fig. 7–14 One–Firefighter Webbing Handcuff Knot Drag (Hervas)

Turnout coat rescue harness

Some turnout gear manufacturers will provide a harness that is integrated into the turnout coat. It is sewn into the coat in such a manner that it will grab the wearer around the chest and under the arms when the loop located near the collar is pulled by a rescuer. This harness system has a number of advantages:

Fig. 7–15 Two–Firefighter Webbing Handcuff Knot Drag

- Very basic and quick to use in difficult fire conditions
- Lightweight, protected, and concealed in the outer shell material
- Can be standard issue to all personnel
- Not costly

Firefighter side-by-side rescue drag method

Two firefighter rescuers perform this rescue method in the event that the firefighter victim is very heavy and difficult to move and requires more than one rescuer to perform a rescue drag. Its advantage is that the rescue can be performed faster and with less effort with two rescuers, benefiting both the rescuers and the victim. In performing the side-by-side rescue drag, the rescue area must allow for the width of the

Fig. 7–16 Side–by–Side Firefighter Rescue Drag (Hervas)

two rescuers and the victim as they move through the building. As the rescuers stay low to the floor, they both simply perform a one firefighter rescue drag from both SCBA shoulder harnesses. The firefighter rescuers become mirror images and work together to pull the victim to safety. If there is difficulty for either of the rescuers in grasping the victim's SCBA shoulder harness, they simply sit the victim up and raise the SCBA from the bottom of the air cylinder to take the weight off the harnesses, then lay the victim back down and perform the rescue drag. The only disadvantage to this technique is the width of the rescue drag ,making it difficult to move down the narrow residential hallways and other close areas.

Firefighter push/pull rescue drag method

In the event there is not enough clearance due to small rooms, narrow hallways, and collapse debris, a "push/pull" rescue drag can be used. This method will allow the two rescuers and the victim to be in an "in-line" position, making it easier to remove the victim through narrow passages. You should remember two important things in order to assure success using this method. First, it is best if the two rescuers position on the same side of the victim to keep a narrow profile and to generate the maximum strength to move the victim. Second, the firefighter rescuer pushing the victim must push on the victim's leg at the hamstring. If the rescuer pushes above the knee, the victim's leg will flex back. The result will be little or no "push," thereby placing more work on the rescuer at the SCBA shoulder harness. The weight of the rescuer trying to push can also shift onto the victim, making the rescue drag much more difficult. In spite of these factors, when done correctly this method has been preferred over other rescue removal drag methods.

Fig. 7–17 Push/Pull Firefighter Rescue Drag (Hervas)

Unconscious Firefighter SCBA Removal

There will be times when a rescue removal of an unconscious firefighter will require the removal of their SCBA. Removal of the SCBA, as performed by the rescuers, will generally have to be done when the firefighter victim is entangled, caught in a very confined space, or must be lifted up to a window. Understanding that the rescuers may have expended a great amount of energy just to get to the victim, the SCBA removal is designed to be as easy as possible on the rescuers by using a roll technique.

Fig. 7-18 Removal of the High SCBA Shoulder Harness (The harness is extended as the firefighter victim's hand is placed over the heart. The arm is then fed through the harness for removal. It must be noted that some brands of SCBA provide chest harnesses that must also be removed.) (Hervas)

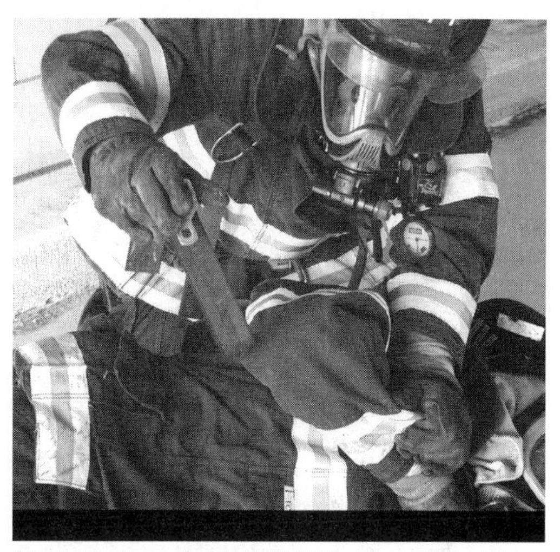

Fig. 7-19 Removal of the SCBA Waist Harness is Next (Once released, if the victim is wearing any type of utility belt, it is advisable to remove it to reduce the chances of entanglement during a rescue drag or ladder rescue.) (Hervas)

Fig. 7-20 If the Victim Is Still Wearing a Connected SCBA Facepiece, the Mask-Mounted Regular or Breathing Tube Must be Removed (Hervas)

Fig. 7-21 The Victim's Lower Arm Is then Fully Extended Over the Head (If this is not done, there is a chance that the victim will be further injured when rolled and the lower SCBA harness will also become entangled under the arm.) (Hervas)

Fig. 7-22 The Victim Is Then Rolled Facedown Leaving the SCBA Behind on the Floor (The SCBA is then pulled upward off of the victim's arm. Once removed, the victim is then rolled back to a face-up position.) (Hervas)

Confined space, unconscious firefighter SCBA removal

If the victim is in a confined space and cannot be rolled out of the SCBA, turn the victim onto the SCBA cylinder, release and remove the shoulder and waist harnesses, and then sit the victim up and remove the SCBA.

Firefighter victim change of direction spin

When rescuing a firefighter victim, the ability not to only move the victim forward, backward, upward, and downward is important but also the ability to change their direction. The firefighter rescuers may need to change the victim's direction of travel for the following reasons:

1. The victim may be lying in the opposite direction of the rescue drag
2. There may be a need to turn the victim to position them under a window for a window rescue
3. The victim may need to be turned around a corner or around other obstacles such as machinery or furniture
4. The victim may need to be disentangled from wire, pivoted off a snagging nail, or moved to avoid fallen debris

To "spin" the firefighter victim, it is important to understand that leaving the upper torso on the floor is best since it is the heaviest part of the body. Raising the victim's legs to at least 90° and then spinning the victim on their back will allow the rescuers to turn the victim with minimal effort.

There are several advantages to using the spin method to reposition the victim. First, the victim can be turned while wearing or not wearing an SCBA. Second, the victim's body length, folded in half as the legs are raised, reduces the chance of the victim becoming entangled in furniture or getting caught on a wall. Finally, the "spin" method can be performed by one or two rescuers and does not require a great amount of effort.

1. Grasp the turnout pants cuffs or the boot heels of the victim. (If a second rescuer is available, then each rescuer will each take a leg and the collar.)
2. Raise the victim's legs to at least 90°.

   Note: As the raised legs are pushed back toward the victim, the weight will transfer onto the victim's shoulders, which will help with turning the victim. Do not let the legs bend at the knees.

3. While holding the legs up, grasp the "grab point" on the victim's turnout coat collar.
4. Turn the victim as needed.

The only additional task is communication if there happen to be two rescuers. One of the rescuers can use a simple verbal command such as, "Head to me," which will indicate the direction of the turn. For example, if the victim only needs to be turned 45°, then the rescuer closest to that point should initiate the command to the other rescuer, "Head to me," and the victim should be turned toward the rescuer calling the command.

More complicated commands as "Turn clockwise," or "Go north," and so on can be easily confused and should not be used, since the result easily can be more confusion and frustration, unnecessary expenditure of energy, and the victim being pulled apart!

Fig. 7-23 As the Firefighter Victim's Legs are Raised, the Collar is Grabbed, and Then the Firefighter Victim is Turned, as Needed, for a Change in Direction (Kolomay)

8

Upper- and Lower- Level and Confined-Space Rescue Methods

In most cases, firefighter rescues that must be executed from a second story or basement will involve a RIT. The level of difficulty is much greater when one has to move an unconscious firefighter up or down staircases, ladders, or with the use of rescue ropes. Although the rescue may be initiated by firefighters who originally were nearest to the firefighter victim(s), their lack of energy, low SCBA air levels, and any sustained injuries will limit their ability to perform any type of rescue. In working with upper and lower level firefighter rescues, the main difficulty that confronts rescuers is trying to move the 250 to 300 pounds of victim weight (this includes the victim's actual weight plus turnout gear, absorbed water, and debris) to safety. Additionally, moving this weight up and down, as opposed to across a floor, is most difficult because the rescue will have to contend with confined spaces, irregularly shaped holes, sloped floors, weakened floors, and staircases. The "4 to 1" rescuer-to-victim ratio is derived from upper and lower level rescues due to the amount of equipment, SCBA air, and strength that might be needed.

Firefighter rescue from an upper or lower level has an increased risk because of the following factors:

- If a firefighter victim has fallen through a weakened portion of a floor, rescuers moving in the direction of the hole can either fall through the same hole or, due to the increased weight (live load), further collapse the floor, taking everyone into the basement.

Fig. 8–1 A Firefighter in Need of Immediate Window Rescue (Hervas)

- Committing rescuers into a basement to locate and rescue a firefighter victim has great risk, because often there is only one way in and only one way out.
- When rescuers come together to perform a rescue lift or drag on a staircase, the live load within a four- to five-step span is at least tripled compared to what the staircase was designed to hold, thus increasing the chances of collapsing the staircase. In addition to tripling the live load, the weight of the rescuers is shifting and impacting as they move up or down the staircase.
- When lowering a firefighter victim from a roof or window and utilizing ground ladders, the rescuers' risks of not having secure ladders and falling are highly increased.
- Since upper and lower level firefighter rescues can become so physically demanding and complex, the amount of time committed to the rescue can dramatically increase the risk of collapse, threatening fire conditions, loss of SCBA air, and injury to the rescuers.

The previous factors are just some of the many increased risks to the RIT members when performing upper- and lower-level firefighter rescues. The rescue methods discussed in this chapter will involve ladders, ropes, and removal skills that work well in a training environment but can also easily go awry in reality if all personnel are not adequately trained. Training should be progressive as firefighters learn the skills step-by-step under very clear and controlled conditions. It is recommended that they should then progress to simulated smoke conditions in actual training structures.

Narrow-Staircase Firefighter Rescues

What qualifies as a "narrow staircase" in the context of firefighter rescue is the inherent lack of space, the inability of the rescuers to position themselves on the sides of the victim, and the difficulty of having more than two rescuers perform the rescue. Narrow staircases are typically found in single- and multiple-family homes. Even though narrow staircases exist in other structures, such as industrial catwalk staircases and high-rise penthouse mechanical rooms, the most common structures where firefighters will be operating on narrow staircases will be in residential buildings. Usually, the width of the staircase will depend on the age of the building. Older construction will have narrower staircases about 30 in. wide and sometimes narrower, whereas staircases in newer construction will average 36 in. wide. An upward rescue on a narrow staircase is one of the most difficult rescues to accomplish for several reasons:

- The upward climb
- The steepness and turns of the staircase
- The lack of a banister for the rescuers to grab or a banister that easily tears out of the wall
- The likelihood of entanglement from metal nose guards catching the victim's SCBA to balusters and catching the turnout gear to banisters
- The rescuers fatiguing quickly because of the upward struggle and limited number of rescuers that can be on the staircase
- The greater chance for collapse during the rescue on some narrow, wood-constructed staircases rather than on wide staircases
- The ability to use only the more difficult head-to-foot rescue due to not being able to position themselves for the easier side-by-side removal method as is possible with a wide staircase

Fig. 8–2 Narrow–Staircase Construction (Firefighter rescuers cannot work side-by-side with full-protective turnout gear and SCBA.) (Hervas)

Narrow-staircase firefighter "push/pull" rescue method

Once again, rescuers must contend with the overall victim weight, which can typically be anywhere from 250 to 300 lbs. The restriction of space in a narrow area will also limit the number of rescuers who can lift and maneuver the victim's weight, causing, in many cases, the victim to be moved almost vertically The push/pull rescue method is designed to rescue an injured or unconscious firefighter victim from a cellar or basement to the floor above. This method can be done on almost any staircase design, but it is especially effective on narrow staircases. This method consists of the following factors:

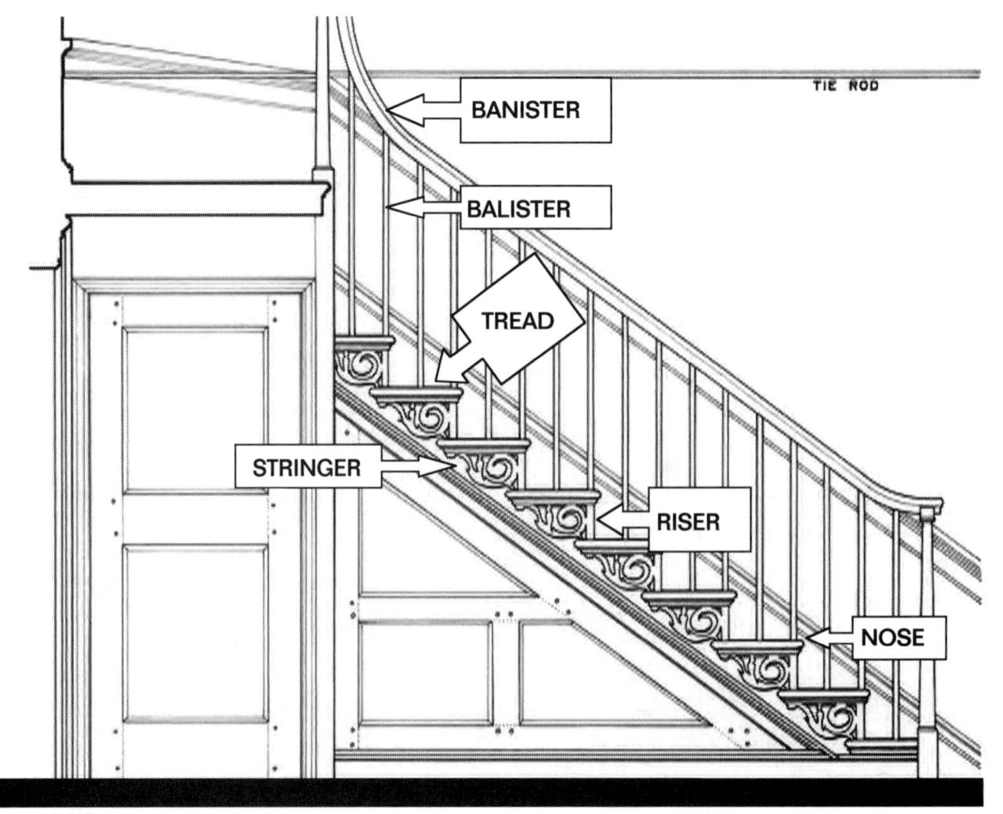

Fig. 8–3 Stair Construction Diagram

- It requires two trained firefighter rescuers.
- The rescue will be conducted in two stages since firefighters generally have difficulty in lifting the firefighter victim from the basement floor up to the steps.
- The method is designed for immediate rescue, as opposed to a more complex rescue operation with haul systems, wall breaching, and use of a rescue basket.

Steps in the push/pull method:

1. The firefighter victim is positioned at the bottom of the staircase and sitting up with their back against the staircase.
2. The rescuer at the head grasps the SCBA shoulder harness. The rescuer may have to tighten down the shoulder harness before grasping. To have the maximum amount of lift off of the steps, the shoulder harness cannot be loose and/or allowed to extend. With some particular brands of SCBA equipment, the excess harness strap material might have to be doubled over and also grabbed.
3. If a banister is available and is durable enough, the rescuer can use it to assist in both balance and pull.
4. The rescuer at the foot separates the victim's legs and grasps under the knees. With heavier victims and/or fatigued rescuers, the quick grasp under the knees may not be very secure for a carry all the way up the staircase. In such cases, the rescuer should consider placing the victim's legs over the rescuers shoulders as described below.

Fig. 8–4 The Firefighter Victim is Seated at the Bottom of the Staircase and Grasped to be Lifted up onto the Staircase (Kolomay)

5. On command of "Ready? Ready. Go!" the victim is lifted approximately to the third step of the staircase
6. The rescuer at the foot then drops down and places the victim's knees over each of the rescuer's shoulders.

- By shifting the weight onto the rescuer's shoulders, the greater strength to lift the victim up the staircase will be coming from the rescuer's legs.
- It will be important for the rescuer to keep the victim's knees as far over the shoulders as possible while climbing the staircase. If the legs start to slide down, the victim's body will sink and start to hit the steps hindering the rescue effort.

7. On command, the victim is lifted again, and carried up the steps.

- Depending on the weight of the victim and/or the fatigue of the rescuers, the victim may not be able to be carried up the staircase to the top. Several quick stops may have to be made to rest, readjust, or even alternate rescuers.
- It is important that the rescuer at the foot does not push the victim and the rescuer at the head. If this happens, the rescuer at the head and the victim will fall to the steps not only stopping the rescue effort but also possibly injuring one of the rescuers.

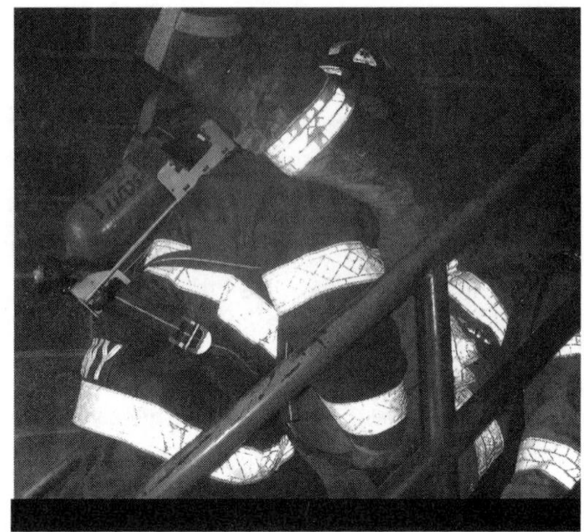

Fig. 8–5 The Rescuer at the Foot Positions the Victim's Knees over the Shoulders (Kolomay)

As is the case of many of the rescue methods, this is a basic method that can be expanded on depending on the many difficulties that may be encountered.

Fig. 8–6 The Rescuers Have Lifted and Now Carry the Victim up the Staircase (Kolomay)

Narrow-staircase firefighter "downward" rescue method

Several years ago, some firefighter rescues down a staircase were being performed in the same manner as rescues going up a staircase. By using the same rescue method going downward as that of going upward, it was found that it took far too much time and effort. The concept of rescuing down a staircase was then changed to include the use of gravity to assist in the effort.

For example, a firefighter may become injured or unconscious in a bedroom on the second floor of a two-story, wooden-frame house. The firefighter is then rescue-dragged out and down the hallway to the top of the narrow staircase. At this point the rescuers need to know things such as, "Which way should the victim come down, headfirst or boot-first?" and "Where and how should the victim be grabbed so the rescuers don't slide or tumble down the staircase?"

Since the firefighter victim is often faceup and headfirst due to the nature of being pulled by the SCBA shoulder harness, the victim should be rescued-dragged down the staircase headfirst. By cradling the victim's head and supporting the neck while pulling from the SCBA shoulder harness, the rescuer rapidly moves the victim and reduces further injury in a controlled escape.

Steps in the narrow staircase downward rescue method:

1. At the top of the staircase, the rescuer reaches into the low SCBA shoulder harness with an "under" grip. The rescuer should not reach through the high SCBA shoulder harness. This rescue method will still work, but if the rescuer reaches through the harness, then they are now "tied" into the victim's harness. If the rescuer must separate from the victim quickly, it will be difficult, if not impossible, to do so.

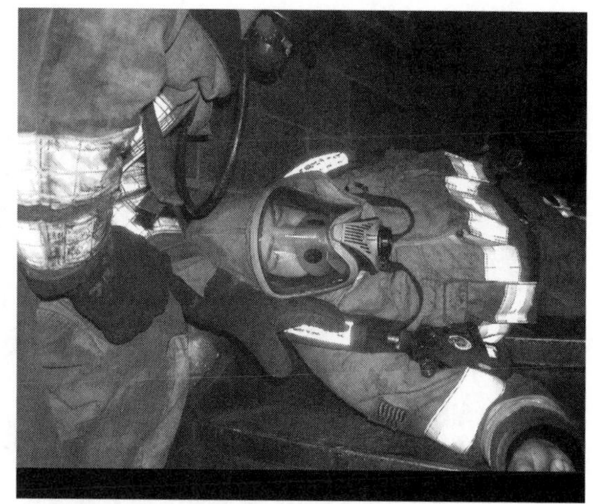

Fig. 8–7 The Rescuer Begins to Reach for the Low SCBA Shoulder Harness from the High Side of the Victim (Kolomay)

2. The rescuer should attempt to insert the gloved hand at the top of the SCBA frame where the SCBA shoulder harness is attached. At that point, there is usually a gap where the rescuer's hand can be easily inserted. The rescuer's hand is then slid down the victim's chest as far as possible. As the rescuer's hand slides up the

shoulder harness, the rescuer's forearm will become tighter and more secure on the victim's neck. This position will help in stabilizing and supporting the victim's head and neck during the descent on the staircase.

3. The rescuer then pulls the victim off of the landing and onto the staircase. It should be noted that if the victim is pulled solely by the SCBA shoulder harness, the rescuers forearm will separate from the victim's head and neck no longer offering protection or support. Therefore, the rescuer should also grasp the victim's SCBA waist harness to pull the victim.

4. To further control the rescue down the staircase, the rescuer must block the victim after each tug with the higher leg to help control the victim's slide, and at the same time check and use each step with the lower leg. Please remember that in the "real world" the

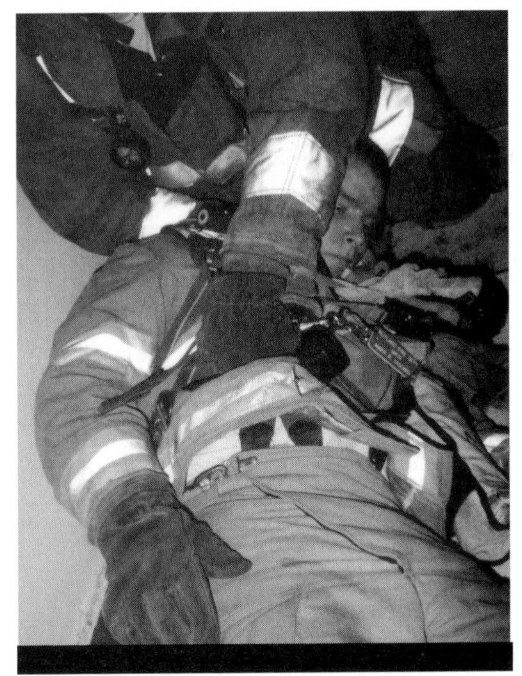

Fig. 8–8 The Rescuer Obtains an Undergrip on the Low Shoulder Harness and Slides the Hand Over the Shoulder as far as Possible (Hervas)

Fig. 8–9 The Rescuer Will Also Grasp the Victim's SCBA Waist Harness to Pull the Victim (Kolomay)

Fig. 8–10 As the Firefighter Victim Slides Down the Staircase, It Will Be Important for the Rescuer to Keep the SCBA Cylinder as Flat to the Stairs as Possible (This will reduce the chances of the SCBA getting hooked and caught on the steps.) (Kolomay)

staircase might be carpeted, waterlogged, and covered with slippery plaster, which is a setup for disaster with one wrong move while going down the staircase. If possible, a second rescuer should be positioned behind the first rescuer to guide the descent.

5. Once at the bottom of the staircase, additional rescuers can grasp the victim's SCBA shoulder harnesses and carry the victim out of the building to medical personnel where he or she can receive proper treatment for burns, etc.

Wide-Staircase Firefighter Rescues

Wide staircases are usually found in occupancies such as commercial, industrial, office, general assembly, and high-rise buildings. They are typically 36 in. or more in width. The most distinct advantage of performing a firefighter rescue on a wide staircase is that four rescuers can become directly involved in the rescue, whereas the narrow-staircase rescue methods are restricted to no more than two rescuers. Typically, rescues involving firefighter victims on wide staircases are somewhat easier than on narrow staircases. In terms of firefighter rescue, a wide staircase will have the following advantages over a narrow staircase:

- The width of a wide staircase usually allows rescuers to position on both sides of the victim, and they have access to both shoulder harnesses, which allows for better pulling of the victim.
- Constructed of steel and/or concrete, wide staircases are usually, structurally, stronger and safer and can hold the live load of up to four rescuers and one victim—a weight in excess of 1,200 lbs in the space of only about five staircase steps.
- Due to the increased room and structural strength of wide staircases, more rescuers can be directly involved in the rescue effort.
- Wide staircases are not, typically, as steep as narrow staircases, thus making an upward drag easier and a downward drag more controlled.
- Wide staircases, usually, have banisters that are securely anchored or welded in place, giving the rescuers a handhold, if needed.

Fig. 8–11 Wide-Staircase Firefighter Push/Pull Rescue (Hervas)

Fig. 8–12 The Four-Point Carry Rescue Method Can Be Conducted Rapidly and with Minimal Workload on the Rescuers Compared to Many of the Other Rescue Methods (Kolomay)

Wide-staircase "push/pull" rescue method

With the advantage of being allowed more rescuers on wide staircases, the push/pull rescue method can be conducted with three rescuers to rescue a firefighter victim from a basement or any other lower level. This method is conducted in exactly the same manner as the narrow staircase, only with the addition of another rescuer who will position at an SCBA shoulder harness to be side-by-side with another rescuer.

Wide-staircase four-point carry rescue method

By once again positioning two rescuers on each of the firefighter victim's SCBA shoulder harnesses, then positioning two rescuers on each of the victim's legs as they grasp under the victim's knees, the victim can be lifted off the floor and up the staircase to safety.

Wide-staircase firefighter "downward slide" rescue method

The "downward slide" requires only two rescuers and, not unlike the narrow-staircase downward rescue method, gravity is to the advantage of the rescuer. This downward slide method will represent somewhat of a belly "seal slide," or can be done in a reverse seated position.

Steps in the wide-staircase downward rescue method:

Fig. 8–13 The Rescuers Are in Position to Use the Wide-Staircase Downward Slide Rescue Method from the Top of the Staircase (Hervas)

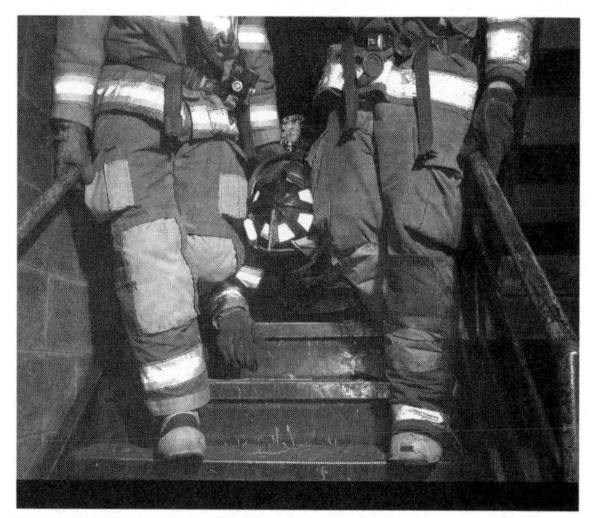

Fig. 8–14 The Rescuers Slide the Firefighter Victim Down the Wide Staircase (Kolomay)

1. Once dragged (headfirst) to the top of the staircase, the victim can be rolled to a facedown position, or the upper torso lifted into an upright seated position.
2. The rescuers stand side-by-side, facing outward toward the base of the staircase, and each rescuer grasps one of the victim's SCBA shoulder harnesses.
3. The rescuers then position their inside legs (closest to the victim) in front of the victim's shoulder. The rescuers' legs will serve as a safety block in the event the victim starts to slide too fast.
4. The firefighter victim is then pulled off the landing onto the staircase and slowly slid downward. When using the seal slide lift method, lift the victim's head high enough to avoid striking the staircase as the rescuers pull upward on the SCBA harnesses.
5. When using the seal slide lift method, as the victim approaches the bottom of the staircase, turn the victim sharply onto the floor. This action will decrease the chance of the victim's head and/or neck being injured if pulled straight into the floor.

Firefighter staircase rescue with a rescue basket

The rescue basket can be a very advantageous and readily accessible piece of equipment for firefighter rescue. Initially used for the transport and staging of RIT tools and equipment, it is also readily available for RIT rescue purposes.

Fig. 8–15 Rescue Basket (Kolomay)

There are various designs of rescue baskets including wire mesh and composite plastics. The composite plastic rescue basket has been a favorite because it slides over uneven surfaces and is less likely to become entangled on debris. The advantages of using a rescue basket for a rescue effort include:

- The victim is packaged and secured allowing for an easier rescue and greater protection from further injury
- The victim can be handled more easily when having to use only two rescuers
- When raising or lowering a victim on ladders or with rope, the victim is more secure, and the rescue can be more quickly accomplished

The main disadvantage of the rescue basket is its bulkiness, which causes difficulty in being able to maneuver around corners, around winding staircases, and through certain windows and doors.

Ground Ladders and Firefighter Rescues

In many cases, fireground tactics have suffered with respect to raising ground ladders around a fire building. Poor tactics, lack of available ladders, poor staffing, and inadequate training are just some of the reasons why ground ladders have not been raised and used as they should have been. Rapid intervention training has been providing new emphasis for the need to aggressively raise ground ladders. Typically, a ground ladder would only be raised at a fire for a civilian window rescue, window ventilation, or to reach a roof. Rapid intervention operations place the emphasis on raising ground ladders around a fire building, not only for routine tactical needs but also for firefighter rescue. The RIT now can provide additional staffing that was once missing on the fireground and is in a good position to proactively raise ladders during fireground operations.

Although it may not be possible to have a ladder raised at every window, interior firefighters should be able to count on finding at least one ladder on any given side of the building where possible. The firefighter in need of a ladder from the second floor could simply get to a window and look in either direction for the nearest ladder. If the ladder is one window over, it will require making it to the next room or moving over to that particular window. If that is impossible because of the fire conditions, at least the ladder can be moved very quickly to the window where the firefighter victims are located. When there is a need for a firefighter to resort to a hang-drop or emergency rappel from a window, it is generally because a ladder could not be raised to that window or even to that side of the building. Often ladders are not raised because of overhead wires, trees, and other obstacles or the more common problem of not having enough firefighters to initially raise the ladders.

With respect to civilian or firefighter rescue, using ground ladders is the least desirable, although the most likely, method to be used. If given a preferred order of choice, the most preferred is the use of staircase, then an aerial device, and last would be a ground ladder. Ground ladders are the least desirable means to evacuate the victim from the building because they require the victim to be lifted and negotiated out of the window. They are not as steady as other means may be, they must be aggressively heeled, and they may have to be set at a steep angle, thus making the rescue more precarious.

When laddering a window, an increasingly accepted fire service standard is the placement of the ground ladder tip just below the windowsill. Whether the ladder is used for ventilation or positioned at the window by the RIT, the ladder heel is then pulled out to adjust the height of the ladder tip to rest just below the windowsill. The positioning to the ground ladder in most cases will be less than 75°. The advantages of this ladder position are as follows:

- The ladder is at a less steep angle, allowing more of the weight to be placed on the ladder by the rescuer carrying the victim.
- The rescuer is typically more secure when able to lean into the ladder, as opposed to balancing on the ladder if a steeper angle were used, while descending with the victim.
- The ladder tip being set just below the windowsill will eliminate it as a potential entanglement problem in the event a firefighter must exit using the emergency

ladder escape method or while negotiating a civilian or firefighter victim out of the window for a ladder rescue.

- Remember Murphy's Law. Even if the ladder tip is only one inch above the windowsill, clothing, rescue belts, or turnout gear can and will get caught!

All things considered, there can be a disadvantage to the ladder being set at an angle less than 75° because it will have a greater chance of "kicking out" at the heel. Of course, on concrete or asphalt the chance of the ladder kicking out will be even greater than when the ladder is set in grass. However, even on grass, the ladder will still require aggressive heeling. During a firefighter rescue, as the rescuer and unconscious firefighter victim begin their descent from the window, there can easily be a live load factor of over 500 lbs at the ladder tip. It is safest for the one or two available firefighters heeling the ladder to position on the outside of the ladder, toe the ladder heel, and brace the closest beam with both hands while looking up to ascertain the stability of the ladder.

Fig. 8–16 Firefighters Aggressively Heel a 24-ft Extension Ladder During a Firefighter Ladder Rescue (Hervas)

It is not recommended to position a firefighter to heel the ladder on the underside while grasping the ladder beams for the following reasons:

- The greatest point of pressure for the ladder to "kick out" is the contact point between the ladder heel and the ground. If the ladder is heeled from underneath, the firefighter's control would be focused on the beams and rungs approximately a quarter of the distance up the ladder not at the contact point between the ladder heel and the ground.
- The firefighter cannot control or stabilize the ladder from the underside in the event it shifts to one side or another.
- The firefighter cannot safely look upward to see the rescue effort and maintain the stability of the ladder.

Fig. 8–17 This Practice of Heeling a Ladder During a Civilian or Firefighter Rescue Is Not Recommended (Hervas)

Firefighter Victim Lifts

Firefighter victim lifts will be necessary when the victim must be lifted over low-level obstacles or up to a windowsill for a window rescue. Firefighter victim lifts can also be very difficult when performed in a confined area, such as after a collapse or maneuvering down a narrow hallway. These situations will require the firefighter rescuers to lift the victim up and over one or more obstacles to continue the rescue drag or carry.

It must be noted that firefighter victim lifts can easily fail in the event the firefighter victim is too heavy for the rescuer and/or the rescuer is too weakened to grasp and lift the victim off the floor. Given a moderate amount of heat in the room, the process of removing an unconscious firefighter's SCBA (if the situation calls for it), completing a firefighter rescue drag, cleaning out the escape window, and lifting a victim who may be waterlogged and covered in plaster can make any rescue effort exceedingly difficult. It will be important for potential rescuers to have practiced many of the details concerning the lift techniques to help assure a successful rescue.

Firefighter window lift and ladder rescue (boots first)

In addition to the RIT rescuers inside the building on fire, the position of the ladder rescuer is of great importance in this type of rescue. From information given to the RIT sector officer or IC, a ladder rescuer on the outside of the building may be able to get into position to perform exterior ventilation (at the rescue window) and make the way for escape much easier for the rescuers and victim.

It is very important that the ladder rescuer have their SCBA turned on and activated in the event there is a need to enter the window or if the smoke and heat conditions are heavy. For instance, if there is a first floor window directly below the rescue window, conditions of heavy smoke and heat rising upward will complicate the rescue requiring the ladder rescuer to use SCBA air.

It must be noted that once the ladder rescuer is in position at the window, there are a number of support tasks that can be done as the firefighter victim is being brought to the window.

- Assure that the bottom window frame is clear of any possible entanglement material such as loose aluminum framework or pieces of glass on which the victim could get entangled. In essence, convert the window into a door.
- Place a hand-held flashlight in the window for the rescuers to target and provide light for the rescue effort.

Fig. 8–18 Firefighter Being Rescued from a Second Floor Window on a Ground Ladder Using the Firefighter Window Lift (Boots First) Method (Hervas)

- Be ready to assist as an extra rescuer inside. In the event the firefighter is too heavy, the rescuers are highly fatigued, or there is more than one victim, an additional rescuer might be necessary.
- Be ready to provide verbal step-by-step direction to the window rescuers due to the fact that they are completely exhausted, cannot communicate, are low on SCBA air, PASS alarms are activated, and the amount of mental stress involved.

With all of the foregoing in mind, let us move onto the window lift and ladder rescue (boots first) scenario. With the knowledge of a weakened staircase, the RIT enters through a rear second floor bedroom window of a bi-level house responding to a Mayday distress call. Once in the bedroom, a PASS alarm can be heard in the hallway. RIT members move toward the victim, and the RIT officer calls for an additional hoseline to cover their rescue effort. Once they find the firefighter victim, the victim is rolled faceup, the PASS alarm is reset, and the RIT officer confirms the Mayday indicating where and who the victim is and that additional help is needed for a ladder rescue. It is found that the victim is no longer wearing an SCBA facepiece. A push/pull rescue drag method is used through the narrow hall and back into the doorway of the bedroom. As the victim is dragged, the rescuers decide to conduct a window lift (boots first). This is to be accomplished as follows:

1. Once in the room, close the door to protect the rescue from any advancing heat, smoke, and fire. Ventilate and attempt to clear as much of the glass and debris from the window as possible. The idea is to "convert a window into a door" by removing window blinds, drapery, glass, and any obstructive framework as quickly as possible. This can be accomplished from the interior or exterior, depending on who's available and if there is a ladder rescuer on the outside.

Fig. 8–19 "Convert a Window into a Door" (Hervas)

2. Once the firefighter victim has been rescue-dragged to the bottom of the window, remove the SCBA from the unconscious firefighter. At this point, the SCBA has served its usefulness as an improvised "rescue harness."

3. Use a firefighter victim "spin" to turn the victim so the victim's legs are vertical against the wall and windowsill. As the victim is spun around, slide the victim into the baseboard, all in one motion. This will place the victim in a seated position against the wall at the windowsill. The closer the victim is seated to the wall, the easier the lift will be. If there is any distance between the victim and the wall, the lift will require both a horizontal move toward the window and a vertical lift up toward the windowsill, thus making the lift more difficult and increasing the chances of failure. It must also be noted that an unconscious victim can be very uncooperative. The legs may have to be supported in the upright position by the window rescuer.

Fig. 8–20 The Unconscious Firefighter Victim Will Be Positioned Against the Wall in a Seated Position (It may be necessary for the ladder rescuer to reach into the window to support the victim's legs during the lift.) (Kolomay)

4. Position one RIT rescuer on each side of the victim. Secure the unconscious victim's legs and arms to prevent them from flailing all over and becoming entangled under the windowsill, under furniture, or even between the rescuers. As a hint, each rescuer can utilize the closest leg shin to the victim's arm to lean in and pin the arms onto the chest.

5. Each rescuer now firmly grasps the bottom of the victim's turnout coat nearest the floor or tight to the victim's waist. Grasping only the outer shell of the turnout coat will allow for a more definite grip. In some cases, if the rescuer grasps all three layers of the turnout coat, the combination of so much material and firefighting gloves result in an unsure grip on the victim.

Fig. 8–21 The Rescuers Utilizing Their Legs to Help Secure the Victim's Arms Before Lifting the Victim (Kolomay)

6. Next, each rescuer firmly grabs onto the victim's turnout coat collar near each ear. These four "grab points" will convert the turnout coat into a rescue sling to be used for lifting the firefighter victim onto the windowsill or, for that matter, over low-level obstacles, such as wooden pallets or fixed-floor electrical boxes.
7. As each rescuer assumes a squatted position, a "Ready? Ready. Go!" is exchanged, and the victim is lifted vertically onto the windowsill and received by the ladder rescuer. Once in a seated position on the windowsill, the firefighter victim may and can become top-heavy, which could result in an unstable position and cause an accidental fall. It is very important that one or both of the rescuers maintain a firm grasp of the victim's collar during and after the lift for control and safety.
8. Once the victim is positioned on the windowsill and while still grasping onto the victim's collar, the window rescuers should move the weight of the victim off the windowsill and onto the ladder and the shoulders of the ladder rescuer. As the ladder rescuer prepares for the victim to exit the window from inside, it will be important for the rescuer to lean in close to the ladder and stay just below the windowsill to avoid being hit by the victim. The ladder rescuer will position each of the victim's knees onto his or her shoulders. As the victim is laid back onto the ladder, most of the victim's weight is also transferred to the ladder.

Fig. 8–22 Both Rescuers Are in Position Side-by-Side in Preparation for the Window Lift (Kolomay)

Fig. 8–23 The Victim is Being Positioned off the Windowsill and onto the Ladder and Ladder Rescuer (Hervas)

9. As the victim's legs are being placed over each shoulder of the ladder rescuer, one at a time, it is important that the rescuer never completely let go of the ladder. As the victim is moved off the windowsill, if there is even a slight hang up and the window rescuers suddenly give an extra, unexpected shove to free the victim, it could cause the ladder rescuer to fall if at least one hand is not holding on to the ladder.

Fig. 8–24 The Firefighter Victim is Brought Down by the Ladder Rescuer One Rung at a Time. (Hervas)

10. The ladder rescuer now descends one rung at a time as the victim is brought down the ladder. The rescuer must step with one foot downward to the rung and must follow with the other foot to the same rung as the first before climbing downward to the next rung. If, at anytime, a rung is skipped and the rescuer's one planted foot slips, it could result in a catastrophic fall for both the rescuer and victim. If at any time the rescuer must reposition the victim while descending the ladder, the rescuer can perform a calf-raise and lean into both the victim and the ladder. Once done, the victim, in most situations, is then stabilized on the ladder, and the rescuer can adjust the victim as needed.

Once the victim has reached the bottom of the ladder, surrounding firefighters will step in and grasp the victim's turnout coat collar and under his or her arms as the ladder rescuer maintains a position at the lower torso. The victim can now be moved to emergency medical services. Concern has been voiced about the victim's head possibly striking each rung during the ladder rescue. This rescue method has shown that:

- When the victim is lifted onto the ladder by the turnout coat and slid down the ladder, the coat tends to ride up the back of the victim, providing some neck support and keeping the head, somewhat, away from the ladder rungs.
- When the descent down the ladder is slow and controlled, it significantly reduces the chance of a serious head injury.
- In the event a second ladder can be raised next to the rescue ladder, the rescuer on the second ladder can assist in protecting the victim's head by providing additional head and neck support.

One-firefighter window lift and ladder rescue (headfirst)

The ladder rescue (headfirst) is not the preferred method of a firefighter victim ladder rescue over other methods, but it is effective and rapid. It can be especially difficult when only one firefighter is available. In the following scenario, two firefighters search the second floor of a two-story ordinary house while the fire extends into the staircase and walls. Suddenly, one of the two firefighters run out of SCBA air, becomes disoriented, and subsequently unconscious from the heavy smoke conditions. The victim's partner, who now becomes a rescuer, resets the PASS alarm, communicates a Mayday distress call via radio, removes the mask-mounted regulator from the SCBA facepiece, and closes the door to the room for protection from additional heat and smoke. The RIT finds that the staircase is too weak to climb to the second floor and notifies all fireground personnel of the problem. The rescuing firefighter must now perform a "one-firefighter rescue drag" with the victim's SCBA to the nearest window. As the rescuer is alone, the victim must somehow be raised to the windowsill for a rapid window rescue. The rescuer will have to perform a "one-firefighter rescue lift and ladder rescue (headfirst)." The first three steps in this procedure are listed below:

1. Ventilate and attempt to clean out as much glass and framework from the sill as possible
2. Remove the victim's SCBA using the unconscious firefighter SCBA removal method
3. Turn the victim 180° so the victim's head is away from the window

The rescuer will then perform the one-firefighter rescue lift.

One-firefighter rescue lift Ladder rescue (headfirst)

The ladder rescuer can also accomplish this method whether the victim is wearing an SCBA or not. In the event that outside rescuers or the RIT locate a firefighter victim positioned on the windowsill, a ground ladder must be raised carefully to the windowsill. The following process is then executed:

1. The ground ladder tip is set and heeled below the windowsill.
2. Once the ladder rescuer reaches the victim, the window rescuer grasps the victim's turnout coat and lifts the victim's legs to assist the victim out the window.

Fig. 8–25 The Firefighter Victim Must Be Sitting Up As the Rescuer Assumes a Weightlifter's Squat Position, Reaches Under the Victim's Arms, and Firmly Grasps the Forearms (Hervas)

Fig. 8–26 Once the Rescuer Has a Locked Grasp Around the Victim, the Rescuer Assumes a Weight Lifter's Squat Position and Lifts the Victim (Hervas)

Fig. 8–27 An Unconscious Firefighter Victim in Position for a Firefighter Ladder Rescue (Headfirst) from a Second Floor Window (Hervas)

Fig. 8–28 The Ladder Rescuer Has Placed an Arm Under the Victim's Neck at the Shoulder as the Victim's Weight Shifts onto the Ladder (This helps ensure the victim is secure on the ladder.)

3. The ladder rescuer receives the victim by placing the first hand under the lower arm of the victim. In certain situations when the victim may be very large and/or slippery, thus making any type of grip by the rescuers difficult, the ladder rescuer should place the first hand in front of the lower shoulder and under the neck. This will assure that the victim will not gain downward momentum and fall off of the ladder

4. Once the victim is securely on the ladder, the window rescuer can stabilize the victim on the ladder and temporarily hang onto the victim, thereby taking some of the weight off the ladder rescuer and allowing the ladder rescuer to change hand position from under the neck to under the lower arm. Communication between the window rescuer and the ladder rescuer is vital at this point. Keep in mind that there is a good chance that the ladder rescuer's SCBA facepiece could be fogged up with moisture or smoke conditions could obscure vision and even the ability to communicate. Make sure you clearly understand each other as you conduct the rescue.

Fig. 8–29 The Hand and Arm Positioning of the Ladder Rescuer Under the Victim's Arm Will Provide a More Secure Hold and Centering of the Victim on the Ladder (Hervas)

5. As the ladder rescuer descends, a majority of the victim's weight should be on the ladder. It

is also very important for the ladder rescuer to descend one rung at a time and maintain control during the descent. Do not descend down every other rung. If at any time the ladder rescuer slips off an icy or wet ladder rung, both the rescuer and the victim can fall to the ground possibly resulting in serious injury.

6. As the ladder rescuer and victim near the ground, it will be important for the firefighters heeling the ladder to lend verbal support indicating the number of rungs remaining before stepping to the ground.

Fig. 8–30 Additional Ladders Raised on Each Side of the Rescue Ladder (It is important that the ladder beams are not so close that the ladder rescuer cannot grasp the beams during the ladder rescue.) (Hervas)

Fig. 8–31 Firefighters at the Bottom of the Ladder in Preparation to Remove the Victim from the Ladder (Hervas)

Compensating for severe ladder angle. If the ladder must be raised in a gangway between two buildings or raised to avoid electrical wires and trees, then the ladder will be at a more severe angle (e.g., 80° angle). That type of angle will place additional weight on the arms of the ladder rescuer, possibly to the extent that the rescue cannot be performed. There are several alternatives for ensuring a safer rescue effort with a more severe ladder angle:

- Raise a second and/or third ladder on each side of the rescue ladder. With an additional rescuer on each side, they will be capable of taking some of the victim's weight from the ladder rescuer and also provide guidance down the ladder. This will not only make the rescue effort safer but also more practical.
- The window rescuer can attach rescue webbing or rope to the victim to reduce the victim's weight on the ladder rescuer from the window during the descent down the ladder.

Lower-Level Firefighter Rescues

Firefighter rescue from a lower grade, confined area, collapse voids, or basement situation can be one of the most difficult rescue efforts to perform. As revealed in a firefighter fatality incident in Columbus, Ohio involving Firefighter John Nance, the difficulty of raising the unconscious weight of a firefighter (approximately 200 to 300 lbs) vertically is extremely difficult and dangerous to both victim and rescuers. The following case study provided a humbling experience that awakened the fire service to the need for improved training in firefighter rescue and survival.

Having been a firefighter for the Columbus (OH) Fire Department[16] for more than 27 years, John Nance, was 51 years of age at the time and was also planning on retiring during the early months of 1988. But on July 25, 1987, those plans changed forever. The 8:00 P.M. start of a 24-hour shift started as any routine shift would have normally started. Station #2, where Nance was assigned, housed 16 firefighters who were assigned to Engine #2, Engine #3, Ladder #1, and Rescue #1. According to the Saturday night tradition, the cook was given a day off, and it was declared "Pizza night!" Dinner arrived around 9:30 P.M., which was later than normal due to a volleyball game. At 10:10 P.M., the tones dispatched Station #2 companies to an old building downtown called the Mithoff Building at 151 North High Street. An arsonist had spread a flammable liquid in the basement and ignited a fire. John Nance, the acting lieutenant on Engine #3, along with driver Marvin Howard and firefighters Tim Cave and Don Weldon departed for the fire scene.

Companies arrived within two minutes and saw smoke coming out of the ground floor of the four-story, 11,500 ft building.[2] As the command post was set in the front of the fire building, the first arriving engine forced open a glass door and advanced a hoseline. As Nance positioned his company in the rear, he then reported heavy smoke coming from a first floor shoe store. As Engine #3's line was charged, Nance, Cave, and Weldon geared up to enter the rear storage room of the shoe store to search for the fire. Following Nance and Engine #3's crew were Ladder #2 and Engine #2, now totaling 10 firefighters in all.

Within about 11 minutes the IC had requested a standby second alarm to fill the staging area with additional equipment and personnel. As the interior companies were having difficulty in locating the fire and the fire conditions were worsening, the IC then upgraded to a second alarm response. Battalion Chief Jerry Lindsay responded on the second alarm and took command of the rear sector. He noticed that the smoke was becoming thicker and was from floor to ceiling. The heavy smoke conditions and the size of the building hampered the interior companies. The firefighters in the front had poor visibility—less than 3 ft with hand-held flashlights. SCBA cylinders now started to run low, and the heat could be felt through the floor indicating it to be a basement fire. Battalion Chief Lindsey had noted during his size-up that the fire conditions were "ominous," giving him a very uncomfortable feeling based on his past experiences with basement fires.

Although the engine company in the front of the building had located a stairway into the basement of the building, the companies in the rear of the building above the shoe store could not find anyway into the basement. Feeling heat coming up through the floor and having used up much of their SCBA air, Nance and his crew left the building and refilled their air. Chief Lindsay stated he saw Nance and his crew start to enter the building for the second time. Lindsay stated that, "I started to talk to John, I said I wanted him to take a rope with him. I wanted him to tie off a rope, a lifeline, and take it in with him." I remember him saying, "Can't we just follow the hoseline in?" I replied, "No, I want you to take a rope." Nance followed the Chief's orders. Shortly after that, Chief Lindsay ordered several other firefighters to take power saws into the building to cut the basement floor for ventilation and to open a vantage point so as to put water on the fire. No one knows how John Nance fell into the shoe store basement, but fire department officials felt he was searching for an area to cut the floor when he fell into a section of weakened floor about 70 ft. from the back door entrance. Firefighter Wilson from Engine #10 had crawled in from the front of the shoe store and had proceeded through an interior door that led into the storage room and had also fallen into a hole grabbing for any possible handhold as he went down. The heat in the hole was intense with heavy smoke and an orange glow could be seen as his legs were being burned causing him to desperately pull himself to safety. As he pulled himself out, he heard Nance screaming for help.

Wilson had noticed that the hole was about 12 ft deep. Wilson stated, "I answered him (John) and then I radioed that we had a man in the basement yelling for help." Communications became very difficult after Wilson transmitted his request for help three times. Companies in the flower shop next to the shoe store did not realize that the basement below had been divided into two separate rooms thereby making accessibility to Nance impossible. By this time, Wilson's air had begun to run out.

At this point Tim Cave left his hoseline and came to the hole with a light. According to Cave, "I found the hole and my arm went down. I asked John if he could see my light. He replied in a very calm manner that he could see it, like he was just standing there waiting for me to get him out of there." At this point, a very important lesson in firefighter rescue was learned as Firefighter Cave extended his hand to grasp the hand of trapped Firefighter Nance. Cave asked Nance, "Can you reach my hand?" Cave then noticed that as Nance reached up to grab his hand, he must have been standing on some stock because the basement was deep. While attempting to pull Nance out, Cave realized that he was slipping into the same hole. At that point Cave told Nance, "I can't pull you out." Nance calmly replied, "OK, give me my hand back."

Inside the building things became confused as the fire conditions became increasingly worse and the other firefighters learned that a firefighter was trapped. Rescue attempts by other firefighters who had reached the hole then commenced. The first strategy was to run a hoseline to provide protection and cooling for Nance, but the hose would not reach and had to be extended. Next, after finding the rope that Nance brought into the building, the rescuers decided to lower it into the hole to pull Nance out. Once Nance grabbed the rope, with the help of three rescuers he was pulled up to within 3 ft of the hole, but then slipped, falling back into the basement. Realizing the amount of weight that had to be hoisted up, an additional call for help around the hole was made. A second rope rescue was then attempted by tying a bowline loop in the line to give Nance a better grip on the rope. Once the rope was lowered, Nance tied two more additional knots in the rope himself. However, even with two additional firefighters on the rope, the rescue attempt still failed as Nance had fallen off of the rope when he was only about halfway up.

Nance was becoming exhausted and low on SCBA air. The suggestion to send a ladder into the basement through the hole was announced. Nance was in full agreement with this suggestion as many ladders were moved into the area of the hole. One of the ladders was lowered into the hole only to find that the hole was too small to allow a firefighter to fit through it with the ladder in place. Once the ladder was pulled out the hole, rescuers worked desperately to enlarge the hole. In the meantime, the fire conditions were quickly worsening throughout the building as well as around the hole. The heat coming from the hole was becoming more intense, and Nance was running out of air, as several firefighters heard him say, "I need air."

While the hole was being opened up and the ladder was being positioned back into the hole, an air bottle was lowered into the basement. Nance then started to climb up to the hole. Unfortunately, he proceeded up the underside of the ladder which resulted in him striking his head several times on the floor joists. As a rescuer reached in and around the ladder while fighting the extreme heat venting out the hole, he tried to pull Nance around to make it through the hole. After repeated efforts, Nance suddenly fell again to the floor. Heat and smoke levels around the hole were becoming intolerable and some visible fire was apparent. Firefighter Wiley attempted another rescue effort by entering the hole, but the hole still was still not large enough for him fit through. After exiting the hole, the hole was again enlarged. Entering the hole again, and taking the hoseline with attempting to knock fire back as he descended, he found an unconscious Nance at the bottom of the ladder. Wiley fought back the fire and grabbed Nance with one hand dragging him with the idea of trying to lift him up the ladder. The heat conditions, the spread of fire, and Nance's weight would not allow Wiley to rescue him from the basement.

Totally exhausted and almost out of SCBA air, Wiley had to get out and allow another rescuer a chance. Firefighter Brining attempted the same type of rescue. Once in the basement, he discovered that Nance's low air SCBA alarm had quit ringing and that he could not find him in the heavy smoke. Brining also had to leave the basement for the same reasons, still without Nance. At this point, Battalion Chief Lindsey approved one last interior attempt for a rescue. However, fire was showing out of the upper floors and the fire conditions had severely changed.

Consequently, all personnel were then ordered out of the building. The fire had gone to four alarms and was finally contained at 5:07 A.M. the next day. Nance's body was recovered from the building Sunday afternoon. The Franklin County coroner's office determined that John Nance had died of asphyxiation with a carbon monoxide level of 64.7%, whereas 6% is enough to cause death.[16] We cannot say for sure what the outcome would have been if the rescuers had been aware of the handcuff knot rescue, but perhaps John Nance may still have been able to retire in 1988 as he had originally planned.

Lower-level handcuff knot rescue

Use of rope in combination with a "handcuff" knot is an option that can be considered using the lower-level handcuff knot rescue method. This method will require one or two rescue ropes no less than 50 ft in length each and the skill of making a handcuff knot as shown.

It is important to note that the handcuff knot requires some minimal practice in training to learn and retain the tying of the knot. Of the few usable "firefighting" knots in the fire service, this is one of the most important. Some fire departments have abandoned the lower-level handcuff knot rescue method because the firefighters could not remember how to tie the handcuff knot. Realizing that firefighters can understand the complicated process of rescuing civilians from an elevator, the hydraulics of pump operations, and be cross-trained as paramedics, then how can the simple process of making the first part of a clove hitch and pulling two sections of the rope through two loops at the same time be too difficult? Let's put it into perspective. Learning and retaining this skill could very well save your life or your partner's life Remember Murphy's Law never fails.

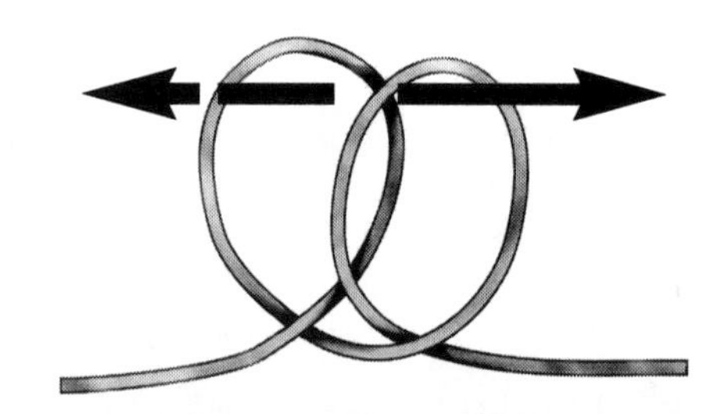

Fig. 8–32 Tying the handcuff Knot (Once two clove-hitch loops are formed, pull the **right inside** loop rope section *through and over* the right loop. Pull the **left inside** loop rope section *through and under* the left loop.)

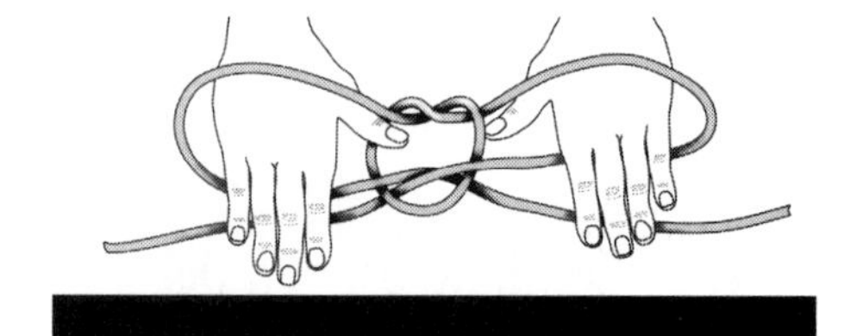

Fig. 8–33 Two Clove–Hitch Loops (Once the inside rope sections are pulled through the loops simultaneously, the handcuff knot is formed and the loops can then be sized.)

Indications for using the lower-level handcuff knot rescue

- A trapped, injured firefighter is conscious but is not ambulatory or capable of moving through debris or up the staircase.
- A trapped unconscious firefighter is positioned in a confined area, collapse void space, or basement without a useable staircase.

Fig. 8–34 Two Sets of Handcuff Knots Affixed to the Firefighter Victim's Wrists with Four Rescue Ropes Extending Vertically Back Through the Hole to Four Rescuers (Hervas)

Contraindications for using the lower-level handcuff knot rescue method

- Potential for collapse in the area of the victim
- Fire and heavy heat conditions not allowing rescuers to use rope at rescue site
- Inadequate training of personnel attempting the rescue
- Inadequate number of rescuers to perform the rescue

Rescue considerations

- Is there danger of secondary collapse?
- What are the fire conditions and the need for hoseline protection?
- Is the building/area of lightweight construction?
- How and why did the firefighter fall through?
- What is the victim's need for SCBA air?
- How stable is the hole? Is bridging or shoring needed?
- Is there strong rescue coordination at the rescue site?
- Can you depend on constant exterior support and size-up?

The lower-level handcuff knot rescue operation

It must be remembered that lightweight construction, structural modifications, and building additions can deceptively lead to additional collapse considerations involving the rescuers. It is imperative that you identify the building construction and if possible, obtain building preplans to provide this information. A disastrous example of how deceptive a building modification can be is seen in a deadly collapse which occurred on October 17, 1966, at 6 East 23rd Street in New York City. The four-story building had first floor store occupancies and apartments above. The Wonder Drug Store had a 98 ft X 15 ft wide basement that had been subdivided by a 4-in. cinder block wall. The wall had been built from floor to ceiling, a distance of about 35 ft from the actual cellar back wall. Firefighters who entered the basement to check for fire were easily deceived into thinking the block wall was the back wall of the cellar. It was not discernible that there was a hidden cellar area behind the wall that had flammable liquids, wood, canvas, and paint supplies from an art dealer stored in it and was well involved with the fire. The basement had a metal fire-retardant ceiling, and the floor of the Wonder Drug Store consisted of heavy, timber floor joists, wooden floor decking, and 5 in. of concrete with asphalt tile.

Since the fire was concentrated and concealed in the cellar, the firefighters in the rear of the drug store did not realize they were standing directly over the fire. The post-fire investigation concluded that the fire had burned for a considerable amount of time in the floor. The brick walls, cinder block partition wall, and combination metal basement ceiling and concrete asphalt floor created a "fire box" condition with extremely high temperatures that let the fire progress for a long period of time, burning away the heavy timber floor joists. When the floor joists failed, 10 firefighters died as they fell into the cellar, and 2 other firefighters died on the first floor just outside of the collapse area.

This incident illustrates that in any rescue operation, the situation can change numerous times within moments, thereby presenting problems for which rescuers might not have been trained. As in the case of this incident, the problem of structural modifications that contradict and hide unseen dangers and pitfalls are why the handcuff knot rescue is a needed skill for all firefighters and especially members of a RIT. Failures in the rescue operation due to Murphy's Law can take over at any time. Structural collapse, the presence of accelerants, and many other problems can plague the rescue effort and necessitate a lower level rescue.

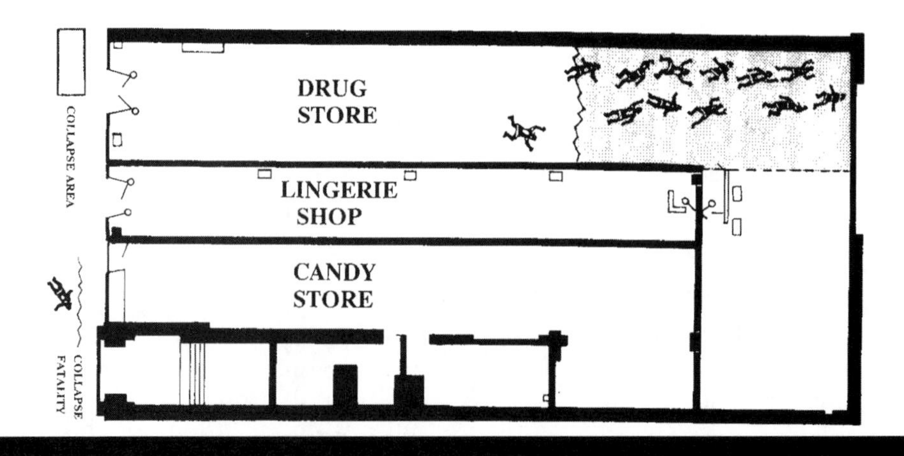

Fig. 8–35 6 East 23rd Street, First Floor Collapse

Additionally, if a firefighter victim weighing 250 lbs fell through a weakened floor decking, then what would happen to four rescuers with equipment who approach that same hole with a combined weight of about 1000 lbs? In addition, once the rescuers start to pull the victim up through the hole, there is potentially another 250 lbs of weight added. Not only could the decking collapse further, but the floor joists may also not be able to handle the load, which could result in a total collapse, costing the lives of many firefighters. Due to all of these factors, the lower-level rescue operation, as presented, involving the handcuff knot, is only a baseline operation to work from, which may require change and/or expansion as the rescue effort evolves.

Fig. 8–36 A Firefighter Victim Being Rescued Using the Lower-Level Handcuff Knot Rescue Method from a Basement (Hervas)

While performing a rescue with this method, do not remove the ladder, even if the rescuers at the top of the hole may feel it necessary to pull the ladder up and out so the victim will fit through the hole when pulled up. Remember, if the rescuers were able to fit going down the hole with the ladder in place, the victim should fit through the hole coming up when pulled up with the rescue ropes. Not only is removing the ladder unnecessary, but it will remove the rescuers' only source of escape in the event something goes wrong. As Murphy's Law has been stated throughout, just as the rescuers make it into the basement, an SCBA low-air alarm will activate, an SCBA malfunction will occur, communications will fail, etc. In addition, the rescuers in the basement may have to contend with darkness and heavy smoke conditions, making the ladder their only reference for an exit. They may also need to tie off a search line to the heel of the ladder to perform a firefighter search operation. The following are the steps needed to perform the lower-level handcuff knot rescue:

1. Assess the structural stability. As the rescuers proceed into the area where the firefighter victim is trapped, constant size-up and evaluation of the building must be done. If the reason the firefighter fell through the floor into the basement is due to a weakened floor from the fire, then extreme caution must be taken as the live load of rescuers and equipment puts additional strain on the floor system.

2. If needed, add some structural stability to the floor system by bridging the floor with ladders, interior doors, wooden stock pallets, or anything that is strong enough to support and distribute the weight of firefighters working in the area.

3. Two hoselines will be needed. Place the first hoseline at the top of the hole to protect the rescuers on top and the second hoseline, with extended hose down in the hole for the rescuers to use for defense against any advancing fire, which would threaten the rescue operation. During a structure fire, any time there is a structural collapse there is a new opportunity for fire to travel. A hole burned through a floor by fire or made from a firefighter falling through is no different from a ventilation hole cut in a roof—it will attract heat, smoke, and fire. Although, initially, the fire conditions may not look pressing, in due time if fire is still burning in the structure, it will try to meet you during the rescue operation much sooner than you expect. On that note, opening an additional ventilation hole in the floor away from the rescue area in an attempt to draw off heat and smoke should be considered.
4. Set up ropes, handcuff knots, and rescuers. (It has been suggested to use different color ropes to differentiate one handcuff knot from another. Although the suggestion has merit in clear and lighted conditions, in real smoke conditions it may serve no functional purpose.)
    - One rescuer shall attempt to identify the middle of the length of rope.
    - Tie a handcuff knot securely at the middle point of the rope(s) before it is sent into the hole. It is important that the handcuff knot is also large enough so the rescuer below can grab and easily slide the loops over the arms of the firefighter victim.
    - The officer in charge of the rescue or RIT will distribute the rescuers around the hole as equally as possible, while guarding them so they, too, do not fall into the hole.
    - It will also be important for the officer to be wary of the floor's stability.
5. Set a ladder into the hole. By setting a ladder into the hole, it will provide a secure means of entrance and exit in and out of the hole. Alternate methods of descending into the hole exist, but they do not always provide for a secure means of escape. If the victim is still conscious, self-rescue up the ladder will be possible, which is the safest and quickest rescue method. In any case, when the ladder is in the hole, one rescuer must be dedicated to securing the ladder from the top of the hole so that it does not slide, fall, or get pulled into the hole. If not secured, rescuers can fall from the ladder, compounding the rescue problem, or the rescuers could quickly lose their escape route to safety.
6. If the victim is unable to self-rescue, no fewer than two rescuers should, as they descend, move a hoseline and equipment down through the hole to the lower

level to provide fire protection and assist in the rescue operation. It is recommended that the following equipment be taken into the hole:

- Portable radio
- Search rope
- Hand-held flashlights
- Hoseline for protection (if needed)
- Thermal imaging camera

7. Perform the needed rescue skills once the firefighter victim is found. The following is a checklist of possible skills you will need:

    - Search for the victim while contending with light or moderate fire conditions
    - Disentangle the firefighter victim from wire or collapse debris
    - Perform a firefighter rescue drag to the base of the ladder
    - Remove the SCBA on the unconscious firefighter victim, while leaving the facepiece on the victim for some respiratory protection. Removal of the SCBA may be necessary to reduce the victim's weight and/or reduce the size of the victim to fit through a small or irregular hole.
    - Share SCBA air with the firefighter victim

8. Although dependent on the fire conditions, it is best for a rescuer to descend the ladder, with the handcuff knot in-hand, to the victim. If smoke conditions are light, the handcuff knot could be tied and lowered to the rescuers below to save effort and time. The handcuff knot rescue method can be performed with one rope and handcuff knot or two of each. The decision to use one or two handcuff knots will depend on how much rope is available, the number of rescuers at the hole, the rescuers' strength, and the size of the victim. If one handcuff knot is used, its success can only be determined at the time the firefighter victim is lifted.
9. Secure the handcuff knot loops to the victim's wrist(s). Each handcuff knot will have two ropes extending up through the hole with a rescuer in position to pull on each of the 2 to 4 ropes. The victim's weight is either halved or quartered depending on whether you use one handcuff or two. (A victim with turnout gear and SCBA weighing 260 lbs will require only 65 lbs of lifting for each person if four rescuers pull the victim up.)
10. Place the firefighter victim against the ladder (if the SCBA is still on), and turn the victim 90° so the SCBA cylinder will not catch on the ladder rungs on the way up.

11. Get underneath the victim to push the victim upward to assist with the pulling. This will be very important when only one handcuff knot is used and as the victim is being pulled through the hole onto the floor above. If a strong push is not possible, there is a good chance the rescue will fail whether there are one or two handcuff knots secured to the victim.
12. Communicate in a coordinated manner as is shown in the following, and raise the victim.

Rescuers in basement: "Down with Knot #1!"

Once the Knot #1 is secured onto the firefighter victim:

Rescuers in basement: "Slack up!" (This will tighten Knot #1 and pull the lines)

If slack must be provided to the rescuers in the basement: "Slack down!"

Rescuers in basement: "Down with Knot #2!"

Rescuers in basement: "Slack up!"

At this point, two handcuff knots are secured on the victim's wrists and the victim is ready to be pulled up through the hole.

Rescuers in basement: "Ready, pull!" (Upon this command, the rescuers above will simultaneously pull on all four ropes, lifting the victim vertically.)

Any rescue operation that involves additional rescuers, equipment, and coordination to rescue a downed firefighter with deteriorating fire and/or structural conditions, is subject to failure. This is because there are more opportunities for failure due to the increased need for communication, chances of injury, fatigue, and many other such factors. The proven way to improve the results of any complex rescue operation is training.

Confined-Space Firefighter Rescues

For a structural firefighter, a confined space can be defined as a narrow channel between two walls (and at times a ceiling) or substantial objects (e.g., heavy machinery), where only one firefighter at a time with full-protective clothing and SCBA can move through. It was not until the death of Denver Firefighter Mark Langvardt that the difficulty of rescuing a firefighter in a narrow confined space became so apparent. At about 2:00 A.M. on September 28, 1992, the communications center of the Denver (CO) Fire Department received a report of a fire at 1625 South Broadway, a two-story commercial occupancy.

The first response consisted of two engine companies, a truck company, and a district chief (Chief #3). Initially arriving companies found heavy smoke showing but no visible fire. The fire building, built in 1980, was a two-story (split-level) building of ordinary construction measuring approximately 50 ft X 60 ft Exterior walls were concrete block, and interior walls consisted of wooden studs and wallboard. The floor joists were laminated, tongue-in-groove, and wooden I-beams. The I-beams were 16 in. thick, the webbing was $^{3}/_{8}$-in. thick plywood, and the cords were each 2-in. X 4-in. Each I-beam spanned the entire width (50 ft) of the building. The building was used as sales offices for several small printing businesses. The upper level contained a compartmentalized maze of numerous rooms, both large and small. The lower level had a similar floor plan with some larger storage rooms.

Engine #16 stretched a hand line to the upper floor, but extreme heat coming from below convinced the officer to reposition onto the lower floor. Members from Truck #16 had then forced entry into two doors (remote from each other) on the lower floor where they had found heavy fire. Engine#16 then knocked down those fires. As additional horizontal ventilation was being performed, another fire was located, now indicating a definite possibility of arson. Truck #16 then requested more ventilation; at 2:22 A.M. Chief #3 called Rescue #1 to assist in PPV and overhaul operations. It was unknown to Chief #3 and members on the lower floor that fire still controlled a large portion of the building on the upper floor as well as the floor and roof assemblies.

Chief #3 then observed fire in the southwest corner of the upper floor. Engine #21, assisted by Truck #16 (Mark Langvardt), then reentered the building to attack the upper floor fire. Personnel found a large body of fire at the southwest corner of the upper floor as smoke and heat conditions intensified significantly. At 2:33 A.M., Chief #3 requested an additional engine and truck, which had arrived at 2:40 A.M. Chief #3 ordered the truck crew to ventilate the roof. The fire on the second floor was intense and stubborn, and progress was slow.

About the time firefighters had located the upper floor fire, Firefighter Langvardt (Truck #16) became separated from his partner and the other interior firefighting teams. At approximately 2:37 A.M., Chief #3 observed what appeared to be a beam of light from a flashlight at a second floor window located just to the left above the building's front entrance. Chief #3 yelled, "Do you need help?" There was no reply. The flashlight was lit momentarily and then disappeared from view. Firefighters outside at the front recognized this as a distress signal. Rescue #1 grabbed hand tools and cutting tools and two portable ladders to make entry into the room through the front, second floor window. Ladders were placed on each side of the window, and Rescue #1 removed the metal window grate and broke the window. Dense smoke, thick enough to obstruct vision in the 20 in. space between the rescuers, poured out of the window. At 2:38 A.M., Chief #3 requested a second alarm.

Fig. 8–37 Recreated Position of Denver (CO) Firefighter Mark Langvardt in the Actual Room (PennWell Publishing, *Fire Engineering*)

The rescue attempt. The initial interior rescue attempt was not successful due to a partial collapse of the upper floor. Hence, Rescue #1 proceeded with the exterior attempt with two firefighters diving headfirst, one after the other, through the narrow window into the room. The drop from the sill was 42 in.. They landed on Langvardt. The victim was facedown, wedged in a fetal position with his head (helmet in place) pressed against the interior portion of the front wall. Rescuers found themselves in an extremely confined space—a 6 ft X 11 ft room filled with filing cabinets and business equipment. The aisle created by this storage was only 28-in. wide. The two rescuers could barely fit in the space together, and they could barely reach into their pockets for small hand tools. There was room for only one firefighter to bend over the victim and try to get leverage to lift him. The standard size door to the room was blocked with equipment, and the only way in or out was a small bi-fold door. This solitary way out was blocked by the floor collapse.

The smoke in the room was so dense that the rescuers could not discern what kind of room they were in, nor could they evaluate Langvardt's apparently unconscious condition. The only thing they could discern was that they could hardly move in this extremely tight place, their brother was down, and they could not raise him up and out of the window. To make matters worse, the advancing fire threatened the rescuers' position even though several hand lines were deployed to push the fire away from the room and the front of the building. Ultimately, numerous rescue attempts were made by rotating rescue teams to remove the trapped firefighter out of the window. All were unsuccessful. Some firefighters thought that Langvardt was stuck or pinned down. This was not the case. The simple, yet tragic fact was that rescuers could not raise Langvardt up and out of the window because of the narrow space and the distance to the sill. One firefighter afterward said that it was like Langvardt "was tied to a thousand pounds of concrete."

Efforts continued as firefighters tried once again to reach the victim from the interior. Access to the victim finally was achieved by breaching through the interior staircase wall. This breaching operation was very difficult because heavy heat and dense smoke had penetrated the stairwell. Working on ground ladders, firefighters used power saws and other cutting tools to breach the wall. Even then, storage shelves, equipment, and stored materials that lined the wall had to be removed. Finally, with the southwest roof of the building in a state of imminent collapse, all firefighters were pulled out of the interior, with the exception of those operating in the rescue room and those who were operating on the foyer/stairwell to keep fire away from the rescue effort. Langvardt was removed from the room at approximately 3:30 A.M., after a 55-minute rescue operation.[17, 18]

One of the main difficulties in this situation was the fact that firefighter rescuers could not position on each side of Firefighter Langvardt to perform a side-by-side lift because of the confined space in which they had to work, an aisle 28 in. that was wide. Raising an unconscious firefighter off the floor to any height from end-to-end is a very difficult task. Add to that the difficult fire conditions, the urgency of the rescue, and the fatigue and frustration of the rescuers.

After this incident in Denver was publicized to the fire service, it became apparent that there was a need to recreate similar situations and devise methods of rescue. In a systematic effort to find solutions, the Denver Fire Department painstakingly recreated the tragedy step by step, documenting every move attempted by the rescuers. As a result of those efforts, a training prop, which conforms to many of the same dimensions that the Denver Fire Department had to contend with, is now commonly used for training for confined space rescues.

Firefighter confined-space rescue method (victim's head at the window)

This particular method does not involve any specialized tools or equipment. It can be considered "down and dirty" using only the firefighters' brute strength, training, and rescue techniques. As was the case in Denver, entry into the building is through a second floor window from a ladder. This method has been proven to be the strongest, most stable, and consistently successful of any of the variations of this rescue.

1. After entry into the window, Rescuer #1 crawls over the firefighter victim then turns 180° to face the window.
2. Next, Rescuer #1 must roll the firefighter victim faceup onto the SCBA cylinder. To rotate the unconscious victim, it is best to turn the victim's shoulders and/or hips to rotate the entire body. (Simply pulling on the victim's arms, legs, or SCBA harness will assist in twisting the firefighter victim into a faceup position as though pulling a rug out from underneath a piece of furniture.)

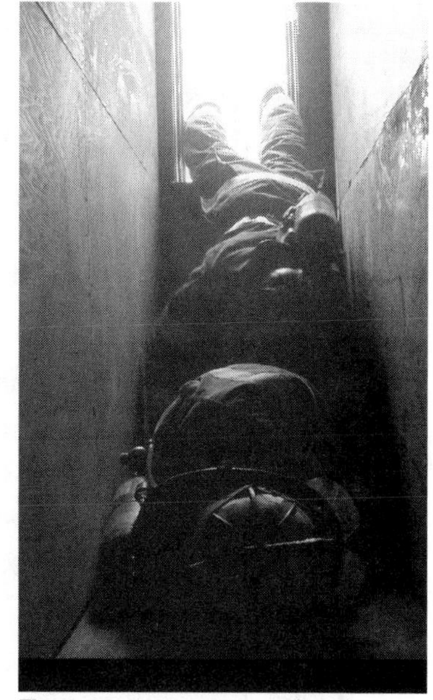

Fig. 8–38 Rescuer #1 Enters Through Window into Confined Space (Hervas)

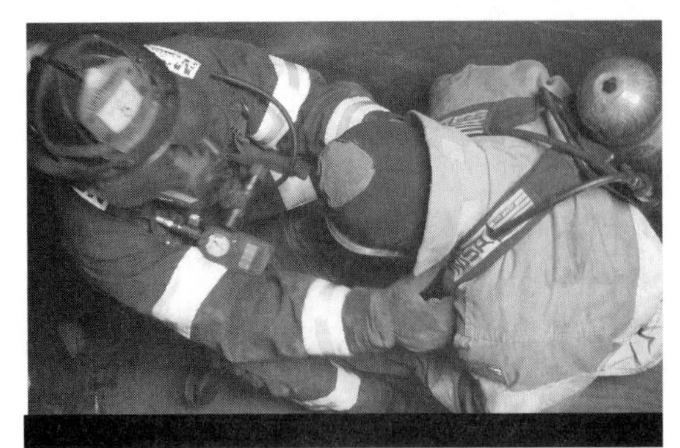

Fig. 8–39 Rescuer #1 Crawls Over Victim, Turns 180°, and Sits the Firefighter Victim Upright (Hervas)

Fig. 8–40 Rescuer #2 Enters Through Window and Assumes a Position Between the Victim Wall Under the Window or Exit Hole (Hervas)

Fig. 8–41 Rescuer #1 Lifts the Firefighter Victim from the SCBA Harness Off the Floor, as Rescuer #2 Lifts from the Victim's SCBA Cylinder (Hervas)

3. Rescuer #1 then pulls on the victim's collar or SCBA shoulder harness to sit the victim upright with his or her back to the window and far enough away from the window to provide space for Rescuer #2 to get into position beneath the window.

4. Rescuer #2 enters the window next and assumes a squat position with their back and SCBA cylinder vertical and flat against the wall under the window. The rescuers feet should be flat on the floor and knees should be bent to form a "chair position" on the floor in order to provide as much support as possible when the victim is eventually lifted and set back down on the knees of Rescuer #2.

5. Rescuer #1 now moves the victim into and between the knees of Rescuer #2 before the first of two lifts. Once situated, Rescuer #1 grasps each of the victim's shoulder harnesses and repositions the grip downward and lower on the victim, near the chest. This will maximize the victim's vertical lift capability.

6. Rescuer #2 then reaches under the victim's SCBA waist harness or air cylinder and will assist with the lift like a forklift.

7. On a coordinated command of "Ready? Ready. Go!" the victim is raised off the floor and onto the knees of Rescuer #2.

8. While the victim is temporarily seated on the knees of Rescuer #2, Rescuer #2 must physically move around to clear the victim's SCBA cylinder to avoid

being struck. Rescuer #2's position is a very precarious one and even more so under fire and collapse conditions.

Fig. 8–42 The Firefighter Victim Should be Seated on the Knees of

9. Once the victim is situated on the knees of Rescuer #2, depending on the amount of confinement in the aisle, Rescuer #1 can either drop to a lower position, placing the victim's knees on each shoulder, or use a "wheelbarrow" grab under each knee.

10. For the second lift, on a coordinated command between the two rescuers, Rescuer #1 will lift the victim's lower torso as Rescuer #2 pushes with both hands (it is important for the hands to be in the formation of a fist for maximum pushing strength and elimination of hand and/or wrist injury) from underneath the victim. Ideally, a firefighter outside the window can reach inward and attempt to assist with the lift and guidance out of the window. The victim will be handed out faceup and headfirst.

Fig. 8–43 Rescuer #1 Can Then Grasp the Knees in a Wheelbarrow Hold or Place the Victim's Knees Over the Shoulders to Lift the Victim out of the Window, Rescuer #2 Will Push Upward, and Firefighters Assisting from the Outside Will Pull and Guide the Victim (Hervas)

11. Once the firefighter victim has been lifted to the windowsill, the victim must be rotated facedown before being moved out of the window. To complete the rescue from the second-story window:

    - An aerial ladder or platform can be positioned next to the window, and the victim can be transferred to a rescue basket and lowered downward.
    - A ladder rescue (headfirst) can be used on the ground ladder(s). The firefighter victim will have to be rotated facedown while on the windowsill.

Fig. 8–44 The Rescuer Will Remove the Victim's SCBA, Sit the Victim Upright, and Use a One-Firefighter Rescue Lift to Raise the Victim to the Windowsill (Hervas)

Firefighter confined-space rescue method (victim's boots at the window)

1. The rescuer enters the window crawling over the firefighter victim to reach the area of the victim's head.
2. The rescuer then turns 180° and performs a confined-space, unconscious firefighter SCBA removal with the victim.
3. Using a one-firefighter rescue lift, the rescuer can raise and set the victim on the windowsill.
4. The firefighter victim is then lowered to the ground by aerial or ground ladder.

Firefighter confined-space rescue with rope assistance

In the event the rescuers are having difficulty because the victim is excessively heavy, difficult to grab, or the rescuers are fatigued, the use of a rope will assist in the lift. By attaching one end of the rope to the victim and running the other end of the rope through a ladder rung above the window and down to the ground, firefighters can pull on the rope when commanded. The pulling force exerted on the rope can almost replace the efforts of one of the two rescuers in the confined space.

Fig. 8–45 Firefighter Confined-Space Rescue with Rope Assistance and a Ladder for a High Anchor Point to Pull (Kevin Tiejl, DeKalb (IL) FD)

9

Firefighter Rescue and Survival Training Methods

In recent years, fire service training has progressed greatly with improved visual aids, computer aided burn facilities, Internet communication, improved research in building construction and fire behavior, comprehensive professional journals and books, and higher training standards. Unfortunately, firefighters have also been seriously injured and killed during training, some as a result of neglect in the area of safety. This chapter has been specifically dedicated to address training and safety in relation to firefighter rescue and survival skills.

Instructor Credibility and Relevance

"Greater love hath no man than this that a man lay down his life for his friends." (John 15:13 KJV) The instruction of firefighter rescue and survival requires the greatest of credibility of the instructors. The life-saving nature of the material to be shared by the instructors and the intensity of the training requires that the attending firefighters be assured of the following instructor qualifications:

- Experience of having operated with a RIT
- Experience of having performed a missing firefighter search and perhaps even a firefighter rescue
- Knowledge gained through several years of firefighting and rescue
- Positive reputation and identification as a leader among peer firefighters

It is recommended that a "team-teaching" technique be used. It has been found that firefighter rescue and survival training is extremely intense and demanding. Not every instructor can always attain the fullest level of credibility demanded for this type of training. Combining instructors into a credible team can round out credibility gaps and provide a high level of instructional quality. One example would be an instructor who is a chief officer and has been in command of a RIT combined with a second instructor who is a company officer and has had the experience of searching for missing firefighters. Chief officers, company officers, firefighters, and emergency medical personnel can all combine into a number of different instructional teams to provided the highest quality of teaching for RITs.

Eight Essential Instruction Points for Firefighter Rescue and Survival Training

With particular attention to the intensity of firefighter rescue and survival training, the following points of instruction, now known as the Essential Eight, are recommended:

1. Relate the training to be conducted to its context at all times.
2. Training must be driven by results not processes
3. All of the objectives must be backed up by criteria.
4. Focus on behavior rather than attitude.
5. Break the training objective and skill into successive stages; don't assume they already know!
6. The learner must perform and understand safety.
7. The training must progressively build and have direction at all times.
8. Firefighter rescue and survival training must be able to measure the results against the goals.

In an effort to better understand each of these Essential Eight instruction points, each point is further discussed below:

1. **Relate the training to be conducted to its context at all times.** The bottom line is, "Can the firefighter relate the firefighter rescue, survival, and RIT training to actual rescues?" Giving the firefighter the ability to put things into context will be the sole responsibility of the training instructors. The mission of a RIT is to search for and rescue missing, lost, or trapped firefighters. The context of that mission will be related repeatedly to actual firefighter fatality incidents and successful RIT deployments where firefighters have been rescued. Each rescue and survival skill must be related to the context of its use in "the street" by an actual incident or example. For instance, when reviewing serious entanglement incidents and the methods of disentanglement, use an actual case study, personal experience, or even a hypothetical situation with the demonstration. Along with some skillful teaching, all of the above will be necessary to drive the point home!

2. **Training must be driven by results not processes.** In the context of firefighter rescue and survival training, although there are numerous skills and methods available, they can all be compared to pieces of a puzzle that must fit together. However, the shape of these puzzle pieces change each time it is put together. In other words, the puzzle, as related to a Mayday distress call, will be different every time. A successful firefighter rescue and the safe return of the rescuers cannot be determined by a canned or scripted training process. In other words, a "black and white" or step-by-step process for rescuing a missing, lost, or trapped firefighter is not realistic. There will be gray areas in every Mayday incident. Such gray areas consist of the uncertainties of just what type of Mayday distress call will be given, how many firefighters are in trouble, the size of the building, fire conditions, collapse conditions, and many other variables along with Murphy's Law from time to time. It will be up to the RIT to pull from training the learned skills and tactics in firefighter rescue and survival to work through such gray areas. Additional help can be gained from a clear and concise standard operating procedure, a planned response of resources in staffing and equipment, and quality training to help compensate for the gray areas.

3. **All of the objectives must be backed up by criteria.** When presenting each of the training objectives, some examples of criteria that must be expanded upon are: behaviors, descriptions, measures, risks, and difficulties. For example, the objective of raising a ground ladder to a windowsill for rescue purposes requires the criteria of how to raise the ladder and the amount of aggressiveness needed when rescuing a firefighter hanging from a window. This criteria in training will "make or break" the objective when it is used in a real rescue. In many cases, instructors teaching firefighter rescue and survival have found that

the criteria for most of the objectives are best learned by being demonstrated. Remember that "a picture is worth a thousand words." How much more so is a demonstration? Much of the criteria to support the objectives involve aggressiveness, passion, focus, and good form, which are all points that are difficult just to "talk" about. The firefighters will respond best when they are given something to imitate.

4. **Focus on behavior rather than attitude.** An opposing force to rapid intervention operations, resulting in some "bad attitudes" among firefighters, is the restriction of the RIT to perform any fire suppression activities. Attitudes come in many different forms for many different reasons, but the end result is to train a firefighter to behave in a responsible manner and to conform to the procedures outlined by their fire department. Firefighter rescue, survival, and RIT training obviously encompasses more than just raw skills. There is a certain degree of emotion. One very important common denominator is that every firefighter has a baseline interest in not dying. The building of rescue and survival skills from that point involves the individual talents of the instructors to gain the respect of the attending firefighters. Behavior is also measurable; attitude is not. It is relatively easy to correct and guide behavior in comparison to someone's attitude. Again, with the realization that dying is not a popular act, it becomes easier to demonstrate and observe the behavior of firefighters learning how to disentangle and drag themselves and their partners to safety with little, if any, "negative attitude."

5. **Break the training objective and skill into successive stages; don't assume they already know!** The training involving firefighter rescue and survival revisits many basic firefighting skills, such as building size-up, communication, SCBA, ladders, ropes, knots, hand tools, and search operations. Unfortunately, not all firefighters keep current with these fundamentals. Then, when introduced to the various skills required in fire rescue and survival that are based on those fundamentals, the training suffers and so does the firefighter being trained. At the risk of being remedial, break down each objective and skill as needed and review it at the basic level, and then move on to the firefighter rescue or survival skill. In most cases, the firefighters being trained will actually appreciate the review, and the training session will be a success.

   The instructors should maintain relevance in their scenarios to foster success for the trainees. It is very important not to lead firefighters into any type of intentional failure at any point in the training program. As an instructor, a "setup" for failure is a "setup" for disaster. If a scenario is designed to demonstrate how difficult a task can be without prior instruction, then, as the firefighters attempt a rescue, the objective will be successful as the instructor explains the objective and goal of the exercise. However, this should not be done with a view to causing the trainees to "fail." Not leading firefighters to failure is even more important later when they are formed into RITs and deployed into training buildings to apply all of their learned skills. Using extremely difficult or unrealistic Maydays in these scenarios will destroy their accumulated learning and could even affect their confidence, thus negating the entire reason for the training program. The training firefighter must correctly understand the objective and criteria the first time they hear it, and they must also correctly learn the skill from the first time they perform it.

6. **The learner must perform and understand safety.** The firefighter, as the learner, must perform many of the actions and skills in firefighter rescue and survival training. We must remember that the goal is to have that training applied when needed most during an actual incident. In certain cases, when a firefighter's life is in danger, the rescue and survival training will have to be second nature and totally reactive to the situation. For example, in the event fire has suddenly cut off the escape route of a firefighter on the second floor of an apartment and high heat is starting to vent out of the window, using the emergency ladder escape method to exit out of the window will have to be as much a part of the person's psyche as breathing. The instructors must understand that each firefighter rescue and survival skill can only vary somewhat in technique before the skill and, consequently, the rescue will fail. Such failures could be related using the following examples. In the event great efforts are put forth in rescuing an unconscious firefighter from a lower level and the handcuff knot is tied incorrectly, then there is a good chance that the victim will fall while being pulled upward. If the rescuer at the foot of a firefighter victim is not secured properly under the victim during a narrow-staircase rescue according to the trained techniques, chances are that the victim and/or rescuer will slip and the rescue will fail.

   The requirement that the learner performs the learned skills is most important, not only for performance reasons but also for safety. As the firefighters learn and perform the various skills, the instructors can observe whether or not they are being performed correctly. Further instruction and correction can be given as needed to ensure that techniques are being performed correctly as ladders are secured and heeled, safety lines are being used correctly, SCBA are buckled as required, etc.

7. **The training must progressively build and have direction at all times.** To many firefighters, the knowledge and practical training of firefighter rescue and survival is new. Because of the amount of information to be absorbed by the learner, it is important that the instructor progressively build step-by-step with a clear direction of where the training is going at all times. While conducting most of the firefighter rescue and survival skill training, it is important that the instructors not only lecture, but also demonstrate skills repeatedly. This method of teaching will prevent an information overload and confusion. As a firefighter

victim is situated on the floor, the instructors will explain the direction and objective of the training, then crawl upward toward the victim and slowly go through the steps necessary to rescue the downed firefighter. Teaching each objective in small portions and then having the firefighters perform the skills in progressive stages has yielded the greatest success in learning and retention.

8. Firefighter rescue and survival training must be able to measure the results against the goals. No matter how extensive the training is, it is imperative that the results of the training be measured against the training goals. As point #7 reinforces, a progressive build-up with firefighter rescue and survival training should lead to the final goals of the training program. Within the context of firefighter rescue and survival training, the results are best measured against the goals with "hands-on" practical training. An example of results being measured against a goal could involve a group of firefighters newly trained in RIT operations. A measurable goal for that group of firefighters would be to operate as a RIT under realistic simulated fire conditions in a training structure and to deploy, search, and rescue a missing, lost, or trapped firefighter in accordance to the training they received. Their success is best measured by the effectiveness of their decisions, execution of skills, chosen RIT tactics, and ability to successfully rescue the firefighter victim based on their training. In the event there is any failure in decision-making, search and rescue skills, or RIT tactics, follow up training can be recommended. Sometimes a simple critique would be sufficient. In more serious cases, additional training might have to be specially scheduled. In either case, the group of trained firefighters will fully understand their strengths and weaknesses within the scope of training they received.

For progressive training and building of firefighter rescue and survival skills, the following five steps can be applied:

FIVE-STEP PROGRESSIVE FIREFIGHTER RESCUE AND SURVIVAL TRAINING PROCESS

1. Explain the context of the training. "On the second floor hallway of an apartment building with moderate smoke and heat conditions, a PASS is heard from an unconscious firefighter."

2. Explain the objective. "We will rescue an unconscious firefighter by performing a one-firefighter rescue drag with SCBA."

3. Simultaneously explain and demonstrate. Simply explain and show the skill to be performed.

4. Reinforce the explanation and demonstration, and answer questions as needed. Observing the body language and facial expressions of the learners and taking note of the questions being asked will determine the number of times or the way in which the skill(s) must be demonstrated.

5. Perform the skill(s). Affirm that the firefighters have, in fact, learned and understood the skill(s) as you observe them perform it.

   The quality of such training is totally in the hands of the instructors. The instructor's credibility, reputation, passion, and ability to motivate are most important. It is important for the attending firefighters to learn from demonstrations where the instructors are fully dressed in protective clothing and SCBA and provide instruction with enthusiasm and energy with the main goal of rescuing one of our own. In many ways to both new and veteran firefighters watching, the RIT instructor becomes a very positive role model.

Preparing Firefighters for Rescue and Survival Training

When introducing the concept of firefighter rescue, survival, and rapid intervention operations, it is has been determined through extensive training that three 8-hour sessions provide the most effective training. The overall training program will consist of approximately 4 hours in the classroom, 4 hours of demonstration, and 16 hours of practical training. It has been found that 30 students should be the maximum class size. Keeping the class to this size ensures that every individual will be able to learn and perform each skill at least once, as well as allowing the instructors to maintain oversight and still deliver quality training. Like any training program, sections can be abbreviated, changed, modified, or even lengthened to adapt to the local fire department's specific needs.

Basic Firefighter Rescue and Survival Training Outline

Session #1: Firefighter Rescue and Survival—Rapid Intervention Operations 8 hours

Session #2: Firefighter Rescue and Survival Skills Training 8 hours

Session #3: Firefighter Rescue and Survival Response 8 hours

From the fire service instructors' perspective, the key aspect to this type of program is the practical training provided for each firefighter. As Chicago (IL) Fire Department District Chief Bennie Crane stated years ago with reference to trainees, "If they can't say it and they can't show it, they don't know it."

Session #1: Firefighter rescue and survival—rapid intervention operations

Session #1 is an 8-hour session requiring approximately 4 hours in the classroom and 4 hours of practical training. As stated in a popular colloquium, "The first impression is the lasting impression." Hence, the initial moments of Session #1 are perhaps the most important. From the reputation of the instructors to the initial contact with the attending firefighters, the importance and seriousness of the material must be conveyed immediately. After introductions and the initial message of the training mission is accomplished, an "icebreaker" in the form of a brief 10–15 minute video of an actual catastrophic mishap on the fireground is shown. In some cases, the mishap may have been elsewhere; in other cases, the video may be from the very town where the class is being given. Regardless, the objective is to remind and capture the attention of the firefighters in class. From that point, it will be important to maintain the attention level by sharing past and recent firefighter injury and death statistics on a national or state basis, as well as any local statistics. Through examples of actual fatality case studies, many of which are available through the United States Fire Administration, the functions, expectations, and demands of a RIT can be covered.

Session #1—Outline for classroom training (approx. 4 hours)

1. Review recent and past firefighter injury and death statistics.
2. Review NFPA 1500 and OSHA 2 In/2 Out.
3. Cover fatality case studies that are most appropriate for the local area.
4. Demonstrate proactive skills and tactics to reduce the need for survival skills or a RIT.
5. Introduce the local operating procedure(s) for rapid intervention operations.
6. Present specific objectives, skills, and examples of rapid intervention operations.

Outline for practical training (approx. 4 hours). It is important for the instructors to give credibility to each skill by relating actual serious injury and fatality incidents to emphasize their importance. However unfortunate, there are many case studies published that will correlate to the skills being taught. For this portion of your training, it is recommended that you review the following subjects:

1. SCBA shared-air emergency methods
2. Minor and serious entanglement emergencies
3. Firefighter rescue with heavy fire conditions
4. Firefighter rescue with light or moderate fire conditions
5. Multiple-firefighter rescuer communication
6. Firefighter rescue drag methods
7. SCBA removal from an unconscious firefighter
8. Tying a handcuff knot

For these activities, the firefighters should be broken into small groups of three to five. Instructors can then simultaneously demonstrate and explain several of the skills while wearing full-protective clothing. Firefighters can then perform the skills while instructors supervise and give assistance as needed. Once those skills are mastered, then additional skills can be taught and eventually combined to evolve into a complete firefighter rescue.

Session #2: Firefighter rescue and survival skills training

Session #2 is a very physically demanding training session. This must be pointed out before any firefighter even starts the training. Anyone attending Session #2 should understand the physical stress that will be placed repeatedly on the shoulders, back, and knees of the trainees in the event any precautions need to be taken. It is also important for the attending firefighters to independently stretch properly and hydrate throughout the training session.

The beginning of Session #2 is designed to instruct trainees in firefighter search methods using basic and wide-area search methods. Search operations are designed to be instructed before the rescue skills are covered since the missing, lost, or trapped firefighter must first be located before any rescue skill can be used. Using search rope, tools, a thermal imaging camera, and a trained plan of action, demonstrate how to search and locate the missing firefighter. When groups of firefighters exceed 15 members in size, it will become necessary to divide the group in half and rotate from skill station to skill station.

Session #2 – Outline of Practical Rescue Skills

1. Demonstrate basic and wide-area search operations with search ropes and the thermal imaging camera.
2. Review and demonstrate (from Session #1):
   - Firefighter rescue with heavy fire conditions
   - Firefighter rescue with light or moderate fire conditions
   - Multiple-firefighter rescuer communication
   - Firefighter rescue drag methods
3. Review and demonstrate aggressive ladder raises for firefighter rescue.
4. Explain, demonstrate, and perform the following methods of rescue:
   - Firefighter victim window lift
   - Ladder rescue (headfirst)
   - Ladder rescue (boots-first)
   - Confined-space firefighter rescue
   - Emergency ladder escape
   - Narrow- and wide-staircase firefighter rescue
   - Lower-level, unconscious firefighter rescue

Training props

For Sessions #2 and #3, it will be necessary to use a training prop, such as an existing training academy structure that is used for live burn training. The advantages are many for these types of training props. First, the training academy structures are very stable and have intrinsic safety devices such as grab railings and engineered anchor points. It is also situated on the training grounds for convenient access to tools, equipment, and apparatus. Second, these structures can easily be converted into simulated buildings with partition walls and furniture for rapid intervention scenario training. Finally, both firefighter rescue skills and scenario training can be conducted during live fire conditions.

These structures usually require some carpentry to frame-out windows and build interior walls. However, depending on the design of the training academy building selected, it can also be difficult to use. Since many of these structures are constructed of concrete and steel, windows, scuttle holes, staircases, and even interior walls are difficult to convert or change for some of the firefighter rescue skills. In some cases, such as the confined-space firefighter rescue skill, it might be more advantageous to construct an independent prop.

Custom training prop

Custom-made firefighter rescue props have been designed to be either stationary and/or mobile and range in cost from $1,000 to $8,000. The advantage of custom props is that the various window openings, staircase designs, and confined-space areas can be accurately constructed for each of the firefighter rescue skills. The training prop can also be set in a controlled area for easier learning and positioning of tools, equipment, and apparatus.

Fig. 9–1 A Training Academy Structure Used for Firefighter Rescue Training (Kolomay)

However, there are two main disadvantages to custom training props. The first disadvantage is a lack of stability. Depending on the carpentry, design, and amount of funding available, the quality and life of the training prop could be in question. The second disadvantage is the lack of reality of an actual structure. Exiting an actual window with a full windowsill or raising a ladder for a window rescue when having to work around real obstacles such as trees and wires is difficult to simulate when using a mobile or stationary training prop.

Fig. 9–2 Mobile Training Prop That Can Be Assembled and Disassembled for Travel (Robert Vonderheide, Illinois Fire Service Institute)

Fig. 9–3 A Stationary Training Prop Located in the Chicago (IL) Fire Department Fire Academy Designed As a Two-Story, Single-Family Dwelling with Engineered Windows, Staircases, and Confined Areas Designed for the Various Firefighter Skills (Kolomay)

Acquired vacant building prop

The use of an acquired vacant building is generally the best option for firefighter rescue training. It will tend to provide the most realistic training, given actual dimensions, heights, and obstacles on which the firefighters can learn and perform the various rescue skills. Selection of the right type of acquired vacant building is important and can save a great deal of preparation work if the overall condition is structurally sound. It can be adequately cleaned out of any debris, and the landscape around the building will allow for realistic laddering to the windows, balconies, and roof.

Fig. 9–4 Acquired Vacant Building Used for Training (Kolomay)

The acquired vacant building is a very realistic prop for the RIT response scenarios. Even with the recommendations of NFPA Standard 1403 Live Burn Training, it is not recommended to use live fires in these buildings during firefighter rescue, survival, and rapid intervention training. In the event there is a problem with the fire attack (e.g., burst hoseline), the team performing the firefighter rescue could be operating above or past the fire and could easily be cut off by the fire or caught in a flashover. The preferred alternative to real fire is to use synthetic water base training smoke. This has proved to be very successful in simulating RIT response scenarios.

Sample Session #2 Training Skill Station Rotation

Whole group	8:00 A.M.	Basic and wide-area search operations with search ropes and the thermal imaging camera
Whole group	9:00 A.M.	Review and demonstrate (from Session #1): firefighter rescue
Whole group	10:00 A.M.	Aggressive ladder raises for firefighter rescue
Group A	10:30 A.M.	Firefighter victim window lift and ladder rescue (headfirst)

Group B	10:30 A.M.	Confined-space firefighter rescue
Group B	11:15 A.M.	Firefighter victim window lift and ladder rescue (headfirst)
Group A	11:15 A.M.	Confined-space firefighter rescue
Whole group	1:00 P.M.	Emergency ladder escape
Group A	2:15 P.M.	Narrow-staircase firefighter rescue
Group B	2:15 P.M.	Wide-staircase firefighter rescue
Group B	3:00 P.M.	Narrow-staircase firefighter rescue
Group A	3:00 P.M.	Wide-staircase firefighter rescue
Whole group	3:45 P.M.	Lower-level, unconscious firefighter rescue

Session #3: Firefighter rescue and survival training scenarios

This final session is designed to be the final application and test of the RIT concepts and skills learned in Session's #1 and #2. Actual teams are formed, search and rescue scenarios are presented to the RITs, and the deployments and rescue operations are evaluated and critiqued.

RIT training scenarios

RIT training scenarios are designed for classroom and practical training. Each scenario addresses different types of buildings, fire conditions, and endangered firefighter incidents. Each scenario will demand a different set of objectives and skills to be achieved to successfully search and rescue the missing, lost, or trapped firefighter(s).

As each scenario is completed, the next is more difficult in an effort to progressively build the training experience and confidence of the firefighter rescuers.

The use of training scenarios beyond the classroom training is essential. Having the firefighters assigned to RITs capable of applying the right skills and tactics, while "thinking on their feet" under pressure, is the ultimate test of efficiency and effectiveness.

In 1996, when a great deal of the firefighter rescue, survival, and rapid intervention training was new to the fire service, many of the rescue skills were taught without further instruction as to how to apply the training to any type of actual firefighter Mayday scenarios. Both authors experienced an eye-opening experience when a group of firefighters, who had been thoroughly trained in firefighter rescue, survival, and rapid intervention operations, were asked to deploy into a smoke-filled training house to rescue two firefighters in the basement. The initial RIT had deployed and soon requested an engine, truck, and squad company for support. The RIT had been told the victims had fallen through a hole in the floor. Between the evening hours and the training smoke, the visibility was realistically low. Typically the lower-level handcuff knot rescue took seven to nine minutes when trained on as an independent skill. What was not factored in was how long the search for the hole would take. As the RIT was searching, their search was slowed for fear of falling into the same hole as the victims. By the time the RIT got to the hole and before the rescue could even begin, their SCBA air was better than half gone. As the PASS alarms continuously sounded and the visibility became increasingly worse, the ladder was brought into the house and placed into the hole. At that point, the initial RIT had two SCBA low air alarms activate. Now the focus was taken off of the rescue of the victims since the two rescuers now had to refill SCBA air and the remaining two could not enter the hole. As the engine company entered with the hoseline, they were ordered to enter the hole first to provide protection, then the squad would follow for the search and rescue. As the engine officer and a firefighter reached the basement floor, they immediately found one of the two victims. It then became the assumption of the squad that the engine was then going to commit to the rescue. As the first handcuff knot was sent into the hole, the engine officer was requesting help from the squad, but communications broke down from the noise and poor radio traffic. As the victim was being tied into the second handcuff knot, the officer on the truck company thought the victim would not fit through the hole, so the officer pulled the ladder up and out of the hole and onto the living room floor. At that point, the engine officer lost contact with the ladder and could not find the location of his partner or the hole. Low SCBA air alarms then started sounding both in the basement and on the first floor, eliminating any possible communication, and the engine officer, the firefighter, and two firefighter victims all "died" in the training scenario.

The main lesson learned from that training experience was that the firefighter rescue, survival, and rapid intervention skills could not be taught independently, thereby ending the formal training. Training scenarios had to be performed in real time, with simulated fire conditions, to effectively complete the training. Even with such training, it is impossible to anticipate every situation that might be encountered. After a thorough post-incident critique, the exact training scenario was attempted again, and it was a successful rescue of both victims. The tactical lessons learned along with the application of the rescue skills made the difference.

Training Scenarios

Training Scenario #1: Search and disentanglement of a conscious firefighter victim

Command skills:

- Identify the Mayday
- LUNAR (Location of victim, Unit victim is assigned to, Name of downed firefighter, Assignment of victim, Radio for communication)
- Assign RIT commander or chief officer
- Deploy RIT operation
- Maintain control of fire

RIT skills:

- RIT single-entry point search operation
- Initial firefighter rescue steps/FRAME (Find the victim, Roll the victim onto the SCBA, Alarm–reset the PASS alarm, Mayday distress call, Exhalation–check SCBA facepiece for exhalation of air)
- Disentanglement of a firefighter victim
- SCBA emergency shared-air rescue method

Objectives:

1. Search the second floor of a single-family dwelling for a missing firefighter using the single-entry point search rescue method.
2. Disentangle the entrapped firefighter
3. Provide emergency SCBA air to the firefighter victim using an SCBA or rapid intervention emergency air system
4. Guide the firefighter victim to safety out of the building

RIT Scenario: During the interior fire attack for a first floor fire in a two-story house, a firefighter on the radio anxiously reports a Mayday from the second floor. The fire-fighter reports being entangled, unable to escape, and is low on SCBA air. The fire is extending through the plumbing walls into the attic, and the smoke conditions are becoming heavy.

Training Scenario #2: Search, disentanglement, and window rescue (boots first) of an unconscious firefighter victim

Command skills:

- Identify the Mayday
- LUNAR
- Assign RIT commander or chief officer
- Deploy RIT operation
- Maintain control of fire

RIT Skills:

- LUNAR
- RIT single-entry point search rescue method
- Initial firefighter rescue steps/FRAME
- Disentanglement of a firefighter victim
- SCBA emergency shared-air rescue method
- Firefighter push/pull rescue drag
- Unconscious firefighter SCBA removal
- Unconscious firefighter "spin"
- Firefighter window lift and ladder rescue (boots first)

Objectives:

1. Search the second floor of a single-family dwelling for a missing firefighter using the single-entry point search rescue method.
2. Disentangle the entrapped firefighter.
3. Provide emergency SCBA air to the firefighter victim using an SCBA or rapid intervention emergency air system.
4. Rescue-drag the firefighter victim to the nearest window and perform a ladder removal.

RIT Scenario: During the interior fire attack for a first floor fire in a two-story house, a firefighter on the radio anxiously reports a Mayday from the second floor. The firefighter reports being entangled, unable to escape, and is low on SCBA air. The fire is extending through the walls into the attic, the staircase to the second floor is weak, and the smoke conditions are becoming heavy.

Scenario #3: Search and rescue of an unconscious firefighter victim who fell into a basement

Command skills:

- Identify the Mayday
- LUNAR
- Assign RIT commander or chief officer
- Deploy RIT operation
- Maintain control of fire

RIT Skills:

- LUNAR
- RIT single- or multiple-entry point search rescue method
- Initial firefighter rescue steps/FRAME
- SCBA emergency shared-air method
- Tie a handcuff knot
- Lower-level handcuff knot rescue operation
- Firefighter rescue drag or Stokes basket removal

Objectives:

1. Search for a missing firefighter who was last reported on the first floor of an ordinary-constructed commercial building. He had fallen through a floor scuttle hole that led into basement (no stairway access to the basement of that occupancy was available).
2. Protect the rescue scene from advancing fire with water and ventilation.
3. Perform a lower level handcuff knot rescue operation.

RIT Scenario: Moderate smoke conditions were found upon arrival of the first due engine company in the rear of a two-story ordinary-constructed commercial. Firefighters begin to search for the fire on the first floor in the end occupancy that happened to be a shoe store. With smoke conditions requiring the firefighters to crawl and heat that could be felt through the floor in the rear, a PASS alarm is heard from the basement. With each company taking an immediate role call, one firefighter from Engine #2 is discovered missing.

Scenario #4: Search and rescue of a lost and unconscious firefighter victim in a large warehouse fire

Command skills:

- Identify the Mayday
- LUNAR
- Assign RIT commander or chief officer
- Deploy RIT operation
- Maintain control of fire

RIT Skills:

- LUNAR
- Multiple-entry point wide-area search operation
- Initial firefighter rescue steps/FRAME
- SCBA emergency shared-air method
- Firefighter rescue drag or Stokes basket removal

Objectives:

1. Conduct a RIT multiple-entry point wide-area search operation.
2. Provide emergency SCBA air to the firefighter victim using an SCBA or rapid intervention emergency air system.
3. Rescue-drag the firefighter victim to the nearest exit point.

RIT Scenario: At approximately 5:00 P.M. on a Wednesday, a fire started by an electrical short near the top of a high-rack storage shelf that is 20 racks high and has ignited cardboard boxes filled with linens. During an interior search for the deep-seated fire that was being suppressed by a sprinkler system, firefighters had to operate in a heavy, cold smoke condition. During the search, a firefighter from Truck #1 became separated between the high-rack storage aisles approximately 120 ft into the building. While reporting being lost and issuing a Mayday distress call via the radio, the firefighter victim was on an SCBA low air alarm. As rescuers attempted to maintain radio communication with the victim, a total loss of contact occurred as the victim's SCBA air became completely depleted. The victim became unconscious and the stand-alone PASS alarm was not turned on.

Scenario #5: Window rescue of an unconscious firefighter victim in a second floor window

Command skills:

- Identify the Mayday
- LUNAR
- Assign RIT commander or chief officer
- Deploy RIT operation
- Maintain control of fire

RIT Skills:

- LUNAR
- Initial firefighter rescue steps/FRAME
- Unconscious firefighter SCBA removal
- Firefighter victim window lift (headfirst)
- Ladder raise to windowsill
- Firefighter ladder rescue (headfirst)

Objectives:

1. Remove the firefighter victim from a heavy smoke condition to the window for immediate rescue.
2. Perform a rescue by carrying the firefighter victim down a ground ladder using the headfirst method.

RIT Scenario: As an officer and firefighter perform a primary search on the second floor of a house in the rear bedroom, a partial roof collapse occurs. The collapse blocks the hallway and bedroom door and also strikes the firefighter resulting in unconsciousness. The officer immediately radios a Mayday distress call stating they need rescue from the second floor window.

Scenario #6: Wide-staircase rescue of an unconscious victim and a conscious victim from the second floor of an apartment building

Command skills:

- Identify the Mayday
- LUNAR
- Assign RIT commander or chief officer
- Deploy RIT operation
- Maintain control of fire

RIT Skills:

- LUNAR
- RIT single- or multiple-entry point search operation
- Initial firefighter rescue steps/FRAME
- Firefighter rescue drag
- Wide-staircase firefighter removal

Objectives:

1. Locate and remove the unconscious firefighter from a narrow hallway.
2. Rescue the unconscious firefighter down a wide staircase.

RIT Scenario: While fighting a fire in a second floor apartment, two firefighters exit the apartment into a heavy smoke condition in the hallway. They are low on SCBA and very fatigued. One of the firefighters suddenly falls unconscious in the hallway from heat exhaustion, and his partner radios a Mayday distress call. As the victim's partner was attempting a rescue-drag, he became too fatigued to continue and remained in the hallway next to the victim.

Scenario #7: Narrow-staircase rescue of two unconscious firefighter victims from a basement fire

Command skills:

- Identify the Mayday
- LUNAR
- Assign RIT commander or chief officer
- Deploy RIT operation
- Maintain control of fire

RIT Skills:

- LUNAR
- RIT single-entry point search operation
- Initial firefighter rescue steps/FRAME
- SCBA emergency shared-air method
- Narrow-staircase rescue operation

Objectives:

1. Locate and rescue-drag the unconscious victims to the staircase.
2. Provide emergency SCBA air to the firefighter victim using an SCBA or rapid intervention emergency air system (if necessary).
3. Perform a narrow-staircase rescue (upward).

RIT Scenario: A fire that had started in the basement of a one-story ranch house had extended through the first floor and through the roof. After the main body of fire had been knocked down, one firefighter with a hoseline in the basement had been working with six other firefighters in overhauling the ceiling and walls. Conditions in the basement started to change slowly when a black ceiling of smoke developed and began banking down with a trace of heat. As the nozzle was positioned near the staircase to the first floor, a firefighter on the opposite side of the basement opened a door to a small utility room that had a developing concealed-space fire. As the firefighter called for the nozzle, the smoke and heat increased, dropping several firefighters to their knees, while causing others to exit up the staircase. The firefighter on the nozzle could not hear the order as the heat was becoming unbearable and the visibility was almost to the floor. The firefighter near the utility room suddenly became disoriented, reporting a Mayday distress call via the radio. Another firefighter took the nozzle and crawled toward the firefighter while directing water on the fire. Upon reaching the disoriented firefighter, they both attempted to follow the hoseline back to the staircase, but both had run out of SCBA air and became unconscious approximately 12 ft from the bottom of the staircase.

Firefighter Rescue and Survival Training Safety Points

1. Do not deviate from the safety points, it can result in death or serious injury.
2. Stay realistic when being innovative.
3. Explain the skill, show the skill, and show the skill again.
4. Know when you are training firefighters and when you are training trainers. There is a difference in the training methodology used for each group.
5. Live victims can be dummies, and dummies can become victims. Be smart, not a kamikaze!
6. Do not use live fire while training in acquired vacant structures.
7. Always survey and warn the firefighters before the training about the extensive physical demands of lifting, carrying, and climbing.
8. Concerning a new firefighter rescue technique, if you talk the talk, then you must walk the walk. Firefighter rescue is one of the most deceptive and dangerous types of training in the fire services.

Epilogue

The Call

They answered the call and raced to the scene
The outcome of this mission could not be foreseen.
It was to them a usual plight
Terror and panic, chaos and fright.
Toward the tower they went with their gear
Pride, courage, and confidence, the badges they wear.
They did not hesitate nor did they balk.
They entered the tower not heeding the talk.

"Look at the firemen!" The people did say
"Don't they know, they're going the wrong way?"
"Be careful!" "Be safe!" and "God bless you!" they heard
No time to reply or say barely a word.
People needed their help, this much they knew
The saving of lives is what they must do.
Not elevators, but stairs they quickly did use
One stair at a time, not a moment to lose.

Helping those in need all along the way
How ironic it was such a beautiful day.
How many lives they saved–we never will know

Onward and upward they continued to go.
Passing countless people on their way to the top
Unwavering, unyielding, unable to stop.
Heroes they're called, but they wouldn't agree
"We're just doing our job," they say modestly.

They continued to climb and climbed out of sight
Trying to climb up just one more flight.
The top of the tower they never did make
For their souls on this day the Lord did take.
Our Brothers were lost, how tragic their fate
But now I believe they are through Heaven's gate.
They now sit around that big kitchen table
Talking and joking, willing, but not able.

Sadness and pain their loved ones do feel
Left with fond memories to help and to heal.
My thoughts and my prayers are with all of you
If we lean on each other, we will make it through.
The stories of these men we'll continue to tell
And take along their spirits are as we answer each bell.
Questions or answers these words are not
Just thoughts of remembrance that I needed to jot.

Brothers we are by the path that we chose
Axes, pike poles, ladders, and hose.
Be safe my Brothers and God bless you all
Because we know we must all answer the call.

Lt. William F. Trezek, Chicago Fire Department

Notes

[1] For more information see Paul Hashagen's *A Distant Fire* (Dover, NH: DMC Associates, Inc., 1995), 194–199.

[2] United States Fire Administration, *Firefighter Casualties,* December 2000.

[3] For more information about the National Institute for Occupational Safety and Health (NIOSH), see *Firefighter Fatality and Prevention Program,* published in February 2001.

[4] For more information on the National Fire Protection Association (NFPA) and the statistics quoted see, *Firefighter Death Rate Not Improved Since 1970's.* Rita Fahy, Ph.D. May 2002.

[5] Illinois Fire Service Institute and Champaign (IL) Fire Department, *Testing Floor Systems,* James Straseske, IFSI Asst. Director and Charles Weber, Captain/Champaign FD, 1986.

[6] Massachusetts Fire Chiefs' Association, *Rose Manor Rooming House–Stoughton, Massachusetts,* (Fire Incident Review Team), January 28, 1995.

[7] Fire Engineering, *New Requirements for PASS Devices,* (Craig Walker and Jack Jarboe), December 2001.

[8] National Institute for Occupational Safety & Health, *Firefighter Fatality Investigation and Prevention,* Case Study 98F-32, 1998.

[9] Incident Review Board, *HIGH-RISE FIRE—750 Adams, City of Memphis, TN,* Office of the Director / Division of Fire Services Inter-Officer-Memorandum Charles E. Smith, Director of Fire Services, 1994.

[10] Columbus Monthly, *The Murder of John Nance,* Columbus Monthly Publishing Corp., Columbus, Ohio. December 1987.

[11] Lieutenant Patrick Lynch, Chicago Fire Department. Interview January 3, 2002.

[12] National Institute for Occupational Safety & Health, *Firefighter Fatality Investigation and Prevention,* Case Study 99F-48, 1999.

[13] Captain William O'Boyle, Chicago Fire Department. Interview May 7, 2002.

[14] Final Report/Southwest Supermarket Fire-38th Avenue and McDowell Road, Phoenix Fire Department. Fire Chief Alan Brunacini. March 12, 2002.

[15] Captain Fred Dimas, Sr., Phoenix Fire Department. Interview, March 18, 2002.

[16] "The Murder of John Nance" *Columbus Monthly,* Columbus Monthly Publishing Corporation, , December, 1987. 40-49.

[17] "Confined Space Claims Denver Firefighter in a Tragic Building Fire," *Fire Engineering.* David McGrail and Jack Rogers. April, 1993, 59.

[18] Captain David McGrail, Denver Fire Department. Interview, June 23, 2002.

Glossary

anchor points. The anchor point is the single secure connection for an anchor. The specific kind of anchor will depend on the specific type of building that is involved.

belay. To keep a firefighter from falling by being attached to a rope. The rope is managed in such a way, or belayed, to keep the firefighter from falling far enough to be injured.

CAD. Computer Aided Dispatch

cold smoke. Smoke that is a by-product of combustion that is not heated due to:

- a smoldering fire
- being spread horizontally over a large area in a warehouse
- rising vertically through elevator shafts in high rise buildings, or
- being cooled by sprinkler systems

fireground. The area or property where the burning structure(s) are located.

flashover. When heat levels rise to a temperature between 1000 and 1200 degrees thereby causing the surfaces of the structure and room contents to simultaneously ignite.

freelancing. When a firefighter acts independently from the group without concern for direction and is unaccountable in relation to personnel location, fireground duties, and communication.

FRAME. Acronym for the initial firefighter rescue steps.

F Find (the firefighter victim)

R Roll (the victim onto the SCBA cylinder if necessary)

A Alarm (reset the PASS)

M Mayday call

E Exhalation of SCBA air

grab points. Various points on a firefighter victim used to grab for rescue drags and lifts.

IFR. Initial Firefighter Rescue

impact load. A load caused by a sudden impact upon part of the building from a structural member dropping to the floor, firefighters stepping up or down stairs, master streams striking a wall, etc.

live load. Any load that is not a permanent part of the structure, such as furniture, utilities, water and snow, and firefighters etc.

LODD. Line of duty death.

LUNAR.

L Locate missing firefighter

U Unit missing firefighter is assigned to

N Name of downed firefighter

A Assignment of firefighter (e.g., roof, rear, outside vent)

R Radio equipped for communication or radio feedback

main wide-area search rope. A larger diameter rope used in wide area search operations as the primary rope into the entry point, and secondary search ropes to be attached to for wall, aisle, and sweep searches.

Mayday. Term for a distress call originally used by ships and aircraft but now applied to any emergency in which a rescue needs to be made.

NFPA. Acronym for the National Fire Protection Association.

NIOSH. National Institute for Occupational Safety and Health

PASS. Acronym for personal alerting safety system.

PPE. See turnout gear.

personal equipment. Equipment that is individually obtained or department issued carried by individual firefighters, as opposed to equipment that is available on apparatus for use by any firefighter.

point-of-no-return. Firefighters who overextend into the building to a point that prevents a safe escape due to:

- Rapidly deteriorating fire conditions causing the firefighters to be caught in a flashover.
- Not having enough SCBA air to make it back to the entry point or any other point of refuge for air.
- Structural failure resulting in collapse.
- Falling contents such as high rack storage, stacked paper bails, and machinery.

RIC. Acronym for rapid intervention crew. The terms RIC and RIT are used interchangeably.

RIT. Acronym for rapid intervention team. The terms RIT and RIC are used interchangeably.

RIT Commander. Officer in charge of rapid and extended firefighter search and rescue operations. The RIT commander is assigned directly to the I/C and positioned at the command post.

RIT Sector Officer. Sector officer in charge of the rapid intervention team operations. The RIT Sector officer is assigned and positioned with the RIT.

SCBA. Self-contained breathing apparatus.

secondary wide-area search ropes. Smaller diameter search rope to be attached to the main wide-area search rope for wall, aisle, and sweep searches.

SOP. Acronym for standard operation procedure(s).

turnout gear (or personal protective equipment-PPE). Protective coat, pants, boots, hood, gloves, and helmet conforming to NFPA and/or OSHA design and safety standards.

violent. A type of injury or death at the site of the emergency as a result of burns, blunt trauma, or asphyxiation.

Bibliography

Brunacini, Alan (Fire Chief). "Final Report–Southwest Supermarket Fire, 35th Avenue and McDowell Road." Phoenix Fire Department, March 12, 2002.

Crane, Bennie L., and Julian L. Williams, Ph.D. "Humanity: Our Common Ground: Your Guide to Thriving in a Diverse Society." Universe Incorporated, September 2000. 22, 35–36, 38–44.

Cull, Frank. "Uniformed Firefighters Association of Greater New York Director of Publications, 23rd Street Fire—A Tribute." Published under the auspices of the Widows' and Children's Fund of the UFA, 1993. 6–22.

Fire Protection Publications. "Essentials of Fire Fighting, International Fire Service Training Association," 4th edition. Oklahoma State University, 1998. 292.

Hudson, Steve, and Tom Vines. "High Angle Rescue Techniques," 2nd edition. Mosby, Inc., St. Louis, MO, 1999.

McCastland, John. "Building Construction." Illinois Fire Service Institute, 1995.

National Institute for Occupational Safety & Health. "Firefighter Fatality Investigation and Prevention, Case Study 98F-05." 1998.

New York State Office of Fire Prevention & Control Academy of Fire Science. "FAST Operations—Firefighter Assist & Search Teams Lesson Plan." March, 1999.

Norman, John. "Fire Officers Handbook of Tactics," 2nd edition. Fire Engineering Books & Videos, 1998.

Richards, Michael. "Chicago Firefighter." The Fireman's Association of Chicago, Spring, 1973.

Russell, Vincent. "Specialized Large Area Search Procedures." Boston Fire Department, Framingham, MA, 2001.

Index

D

E

F

G

H

I–K

L

M

N

O

P–Q

R

S

T

U

V

W–Z

PennWell

Fire Engineering